FUTURE TECH,

RIGHT NOW

X-RAY VISION

MIND CONTROL

AND OTHER AMAZING STUFF FROM

 sourcebooks

Published by Sourcebooks, Inc.
P.O. Box 4410, Naperville, Illinois 60567-4410
(630) 961-3900
Fax: (630) 961-2168
www.sourcebooks.com

Library of Congress Cataloging-in-Publication Data

Future tech, right now : X-ray vision, mind control, and other amazing stuff from
tomorrow / HowStuffWorks.
    pages cm
  Includes bibliographical references.
  1. Technological innovations--Popular works. I. HowStuffWorks, Inc., editor.
  T173.8.F89 2014
  600--dc23

                                    2014018615

          Printed and bound in the United States of America.
             VP 10 9 8 7 6 5 4 3 2 1

Also by *HowStuffWorks*

Stuff You Missed in History Class: A Guide to
History's Biggest Myths, Mysteries, and Marvels

The Science of Superheroes and Space Warriors:
Lightsabers, Batmobiles, Kryptonite, and More!

The Real Science of Sex Appeal: Why We
Love, Lust, and Long for Each Other

CONTENTS

PART II: POWER PLAYS: COOL WAYS TO ENERGIZE OUR FUTURE   80

PART III: HOME IS WHERE THE ROBOT IS: THE INTELLIGENT FUTURE OF OUR HOUSES, HIGHWAYS, AND HOVERCRAFTS   126

INTRODUCTION

F lying cars! Teleporting! Robot servants! Mind reading! Wouldn't you love to have or do any or all of these? Well, you're in luck, because they may actually be closer to reality than you think. These are just a few of the amazing and astonishing things we'll explore in this book, as we peer into what the future holds for science and for us. In *Future Tech, Right Now*, we, the staff at HowStuffWorks, explore some of the coolest (and craziest) ideas and inventions that will have a major impact on our lives in coming decades—from mind control, energy, and drugs that could make you smarter, to intelligent homes and cars, textbooks that can talk to you, and even robotic teammates!

Sound a little far-fetched? We promise we're not making this stuff up. In fact, there is an emerging academic field, futurology, dedicated to studying, analyzing, and forecasting the future in a quantifiable and scientific way. No joke. And that's where we'll begin our investigation. After all, we figured we should consult the real experts on what's ahead: futurists. We've also packed in lots of fun facts, trivia tidbits, and sidebars—just to make sure you know what's in store for you. So strap yourselves in, and get ready to rock and roll!

1

MODERN-DAY FORTUNE-TELLERS:
HOW FUTUROLOGY AND FUTURISTS WORK

People often think futurists are modern-day fortune-tellers who peer into cutting-edge crystal balls. Far from it. A futurist is an educated individual who, after much research and analysis, makes calculated projections about the future—everything from shifting demographic patterns and new technologies to potential disease outbreaks and social conditions.

That may sound a bit more boring than using a crystal ball, but it's a heck of a lot more accurate. Most projections made by futurists are about things that will occur in the next five to fifty years, because it's easiest to make accurate projections involving demographics in the next half-century versus assessing health and social conditions a century out.

Futurist thinking began around the years 1600 to 1800—the Age of Enlightenment. With the publication of Isaac Newton's landmark scientific work *Philosophiæ Naturalis Principia Mathematica* in 1687, people started to realize there was a lot of validity to reason, empiricism, and science. But it wasn't until the nineteenth century that the term "futurist" was first documented in the English language in a work by George Stanley Faber, who was referring to Christian scriptural futurists.

Futurists and futurism really got a kick-start and started gaining popular interest in the early twentieth century with the advent of science fiction. In fact, many consider author H. G. Wells to be the first futurist of the modern era, with other sci-fi authors such as Isaac Asimov and Arthur C. Clarke deemed important early futurists.

Today, futurism is a bona fide career field. Colleges and universities are beginning to create courses and even degrees in futurism. Most talented, professional futurists have multidisciplinary backgrounds, rather than a specific degree in futurism. Anyone considering a career as a futurist must be an inquisitive and creative critical thinker. These individuals also need to be able to imagine all sorts of unusual occurrences and situations, all within the context of solid research.

ⅢⅢ SO WHAT EXACTLY DOES A FUTURIST DO AND WHERE?

Think you'd make a good futurist? There are actually many different types of futurist jobs. You might find one working for the government, nonprofit organizations, or corporations. In the latter case, jobs are often with major marketing and advertising groups, and the positions are called marketing specialists, ideation specialists, or directors of future research. Futurists can also work independently as consultants. Groups as varied as Hallmark, IBM, British Telecom, and the FBI all work with futurists.

In the world of academia, futurists generally focus their efforts on social criticism. Professional futurists who work for companies, whether as consultants or paid staff, try to anticipate market changes and the public's mood to help their clients make prudent business plans. Regardless of the concentration of study, the job involves reading a wide variety of materials related to the subject matter, including media reports and information in statistical databases.

It's also helpful to interview experts in the field, plus other futurists and even members of the public. In addition, futurists spend a great deal of time thinking, analyzing, and strategizing—for example, if x is the current reality, and y

and z are the trends, what might the next logical consequences be? A talented futurist is able to recognize connections in myriad scattered pieces of information.

Because the field is still relatively new, and because futurists' job titles can vary widely, the U.S. Bureau of Labor Statistics doesn't track futurist wages and employment rates. However, it estimated in 2009 that there are five hundred to one thousand professional futurists working in the United States. As a sign of the profession's growth, professional organizations are being created as well, such as the Association of Professional Futurists, the World Future Society, and the World Future Council.

All of this is promising because many people see the rise of futurists as critical to our future. Why? While humans are more than capable of realizing the implications of their actions—pollution, energy consumption, and overpopulation, to name a few—we have a fairly dismal track record at doing what's best for society in the long run. Perhaps if more futurists are around, they'll help us better understand what we should be doing today to ensure a better tomorrow. As you read on, you'll see that many of the craziest, coolest, and most plausible future developments have started as the brainchildren of futurists. Perhaps you'll even come up with a few futuristic ideas of your own!

FIVE FUTURIST PREDICTIONS IN THE WORLD OF TECHNOLOGY

So now that you've gotten a crash course on futurists, let's take a look at some of their latest predictions. Some may seem crazy, but just remember that a generation ago, no one dreamed of smartphones that could talk to you or guide you through traffic, although they're indispensable to us now.

1. MOON, MARS, MORE?

Space exploration has taken some hits in the twenty-first century, with cuts to the U.S. and other international space program budgets. But with the *Curiosity* rover on Mars as of August 2012 and plans to launch the "most powerful rocket in history," the Space Launch System (SLS), by 2017, the National Aeronautics and Space Administration (NASA) is still very much in the business of the future. After the planned, unmanned send-off of the SLS in 2017, NASA intends to send a crew of up to four astronauts into space by 2021. This could be a return to the moon, with the possibility of missions to other planets.

Even with the world economic downturns of this century, individuals and corporations in the private sector also plan to keep aiming for the stars and enabling nonastronauts to explore the universe. Some futurists of decades past would be surprised to see that space travel for every man isn't commonplace, but for a few wealthy adventurers, it's no longer the stuff of science fiction. Maybe their trips will help drive down costs for the rest of us.

GOING UP?

Instead of choosing a floor when getting on an elevator, imagine choosing a planet. Many futurists support the development of a space elevator for transporting people from Earth to the moon and to Mars. Such a planetary lift is probably at least four or five decades away, but the vision is very much alive now as innovators talk of a 62,137-mile (100,000-kilometer) ribbon that can be firmly attached to our home planet and extended to an anchor station in outer space. We will simply ascend on a high-tech thread into the deep beyond and hope not to get stuck between floors.

||||| 2. NEUROHACKING

Will there be a day when you say, "I can't read your mind, you know!" and the reply will be, "Oh, stop it—of course you can!"? It could happen. Neuroscientists are finding ways to read people's minds with machines, and although this has been in the works for decades, real progress is being made by researchers at the University of California, Berkeley, and elsewhere. For example, translating electrical activity from the brain by means of decoding brain waves is one way to help people with dementia; they have complications with neurotransmitters relaying thoughts into comprehensible speech or holding thoughts long enough to voice them before they're forgotten.

On the other hand, it is more than a little frightening to know that science and machines could soon have access to our innermost thoughts. Implications for neurohacking into people's thoughts also have been studied in relation to neuromarketing, which targets people's brains by manipulating their wants and desires through marketing and advertising. Our thoughts and actions could potentially be hijacked by a form of media that makes us think we're getting what we want, when really we're going for something our brains may only think we want.

THE MENACE OF MASS DATA

Even if scientists and marketers can't get access to our brains for neurohacking or neuromarketing, can they get access to our data? More importantly, what will they do with it? With unprecedented amounts of data available about us online, the media, government regulatory agencies, and marketers work around the clock to mine information about us, our habits, and even our relationships. Through narrowed search results, our reading and research have become "optimized" based on what keywords we use, also narrowing our choices in buying products and accessing news and information.

Eventually, data and the machines and algorithms used to manage and make sense of it could largely replace independent decision making. In fact, it is happening at such a speed that it's sometimes hard to remember the data isn't in control. People still control the data, but just who has this control and what they do with it will become an ongoing challenge in the coming decades.

3. NANOTECHNOLOGY, NANOMEDICINE

Technology at the nano level, or nanotechnology, allows for unbelievable precision and a way to copy the work of nature at its most basic functioning, but just how small is a nano? According to the National Nanotechnology Initiative, a sheet of paper is 100,000 nanometers thick, and there are 25.4 million nanos in one inch. A nanometer is one-billionth of a meter!

How is this affecting technology and the future? Nanotechnology is being used for innovations in engineering, medical devices, imaging, computing, and many other fields. Nanomedicine is one area experiencing rapid and dramatic growth. Because many illnesses and disorders in the body take place at the cellular level and grow as ruled by the formation of genetic makeup, nanotechnology has the capability to treat at the very root of the condition, rather than after it has fully spread throughout the body.

This approach can be both preventive and curative because treatment reaches the narrowest and most minuscule centers of control. Neurosurgery and gene therapy are just two areas within nanomedicine that are particularly well-suited for nano tools and technology.

NANOFACTORIES

Taking nanotechnology from an idea to reality means being able to make some very, very fine and small-scaled tools. Nano tools have to be assembled at the molecular level to be tiny enough to perform work at the nano scale, and often, the work of nanotechnology is so specialized that the tools need to be modeled and made specifically for each job. Handling the tools involves careful and minute planning, as well, because of their delicate balance and scale. In the future, those skilled in molecular nanotechnology will be in high demand in the workforce.

4. DARK NETWORKS

As the world gets smaller by sharing cyberspace and social tools, we are becoming a bigger collective target for the bad guys. While our data puts us all "out there" in many ways, that same data enables those involved in dark networks or illegal online activities to take on false, covert identities to plan bigger cyberattacks.

Anonymous is one such group involved in "hacktivism," having wormed its way into sensitive stores of information from the likes of the FBI, Visa, and MasterCard, as well as government websites from the United Kingdom to China, and caused large-scale, disabling computer terror. The group functions as a collective of many individuals and spreads its log-in and computer activities thin enough to lead authorities in too many directions to track. Its attacks target everything from politics to commerce.

As incidents of cyberattacks—and even infrastructure attacks on water systems and electrical grids—grow, billions of dollars are stolen and billions of people are at risk each year. This may lead to increased cyber insecurity, or widespread fear of the very technology people need to go about everyday commerce and communication, as well as many more resources devoted to stopping or preventing such attacks.

5. AVATARS, SURROGATES, ROBOTICS

Maybe you aren't comfortable with all the futurist predictions and even the current rate of technological advance, and that's okay. You can be yourself and interact in the world in a fairly low-tech way while allowing a surrogate, avatar, or robot to live your online and tech life for you. Even the U.S. Defense Advanced Research Projects Agency (DARPA) has budgeted millions of dollars to create avatars that will act as surrogates for real, live soldiers.

While avatars and surrogates were once the stuff of games, virtual reality, and computer interfacing, they are taking on more and more active roles as replacements for living, breathing humans. Or, are they enhancements for humans?

Fully realized robotic machines have become more and more widespread in medical technology and scientific development, both in labs and in hospitals, enabling those with paralysis to move limbs, for instance. "Living" life with "second life" surrogates is likely to become more common every day for those of us in less specialized fields, too.

COGNITIVE CRAZINESS: HOW COGNITIVE TECHNOLOGY WILL CHANGE OUR LIVES

Now that you've learned about some of the predictions futurists are making, let's look at how the future is playing out right in front of our eyes through an interesting field called cognitive technology. First of all, what exactly is cognitive technology?

These days, it's hard not to bump into people who are so absorbed in their smartphones and tablet computers that they're almost one with their gadgets, oblivious to the world around them. This may conjure up vaguely creepy images of the *Matrix* movies, with humans wired into and entirely dependent on an illusion that takes the place of reality.

But that's taking the negative view. To neuroscientists, psychologists, and researchers in the field of artificial intelligence—that is, teaching computers how to mimic and even improve upon the human thinking process—machines positively influence our lives, too. Scientists have come up with the term "cognitive technology" to describe how electronic devices and other tools, from pharmaceuticals to brain-training games, can assist and influence humans' mental activities, such as learning, retaining, and retrieving information from memory and problem solving.

These things won't necessarily do the thinking for us "cognizers," as researchers call those of us made of flesh and blood. Instead, they will give us an added edge over the nonaugmented brain. As cognitive technology researchers Itiel Dror and Stevan Harnad explain: "Cognizers can off-load some of their cognitive functions onto cognitive technology, thereby extending their performance capacity beyond the limits of their own brain power."

As today's technology gives way to devices with vastly more computing power and communications bandwidth, and new generations of psychoactive drugs and electronic implants eventually emerge, cognitive technology is likely to really, really rock our world. In the next six sections, we'll look at some of the biggest areas in which our lives will be affected over the coming decades and centuries: education, entertainment, medicine, the medical profession, our minds, and our computer-based culture.

EDUCATION: HOW WILL FUTURE TECHNOLOGY CHANGE THE CLASSROOM?

Technology used to be a curiosity in the classroom. Computers were scarce and mostly used to teach keyboard skills. But now schools aspire to be the vanguards of the latest technological and educational resources. How is the rapid technological change affecting the way classes are taught, students are engaged, and teachers are using material?

Before we get into some of the cool, new technology that could become an everyday part of the educational system, we need to note that not every classroom has the means to take advantage of the technology of the future—or even the present. Many of the resources discussed in this section likely will be available only to the cutting-edge educational systems and the school districts that have funds to support them.

TECHIE TRENDS IN TEACHING

To see the future of technology in the classroom—and how it will change the way classes look—we should start with college. Although we can't predict how technology will change, we can make some assumptions about how it will trickle down. Of course, university students are now almost uniformly armed with laptops. But they also often work with online learning management systems like Blackboard to post papers, receive instructions, or discuss assignments or lectures. These systems are already starting to show up in lower grades as well, not just so teachers can monitor schoolwork, but also so parents can keep up and keep tabs on their children's grades, homework, and academic progress.

Use of learning management systems speaks to a larger trend that technology might lead to in the future. Customizing the student's learning experience has become a hot (and debated) topic. Harvard University professors Chris Dede and John Richards propose a digital teaching platform called Time to Know that allows teachers to facilitate large- and small-group learning, as well as individual education.

They envision a classroom where each student has a computer, but the teacher can press a button to make all devices freeze, capturing a large group's attention. Beyond that, the teachers would use the broader big-group lessons to let each child figure out how that lesson impacts him or her personally. But how would that occur?

As Dede and Richards point out, our classrooms don't lack content. With the Internet and the technology that lets us connect to resources around the world, what teachers need is a system to assess what content is going to be most valuable to their classes. These digital teaching platforms will have assessments that will theoretically help teachers determine in real time a curriculum that's best for their community, classroom, and even each individual student.

This idea of customization through digital teaching platforms is interesting, but what's even cooler is that textbooks may soon be "intelligent teachers" themselves!

TEXTBOOKS REALLY WILL BE "INTELLIGENT"

Perhaps you remember sitting in study hall, struggling through some thick textbook full of arcane terminology and complicated new ideas that frazzled your neurons to the point of exhaustion. Well, students of the future are likely to have it a lot easier, because digital books equipped with artificial intelligence capabilities will guide them along with the patience and perceptiveness of their favorite kindly professors.

Take the newly developed Inquire intelligent biology textbook for the iPad. It allows students to stop and type in a question like "What does a protein do?" and then presents them with a page full of information specific to whatever concept they're stuck on. The book's intelligent software also contains a machine-readable map that can link any of the five thousand concepts in the text to any other concept and then explore how the two are interrelated. In a study conducted at a California college, students who used Inquire scored an entire letter grade higher, on average, than comparison groups.

There's also been a fair amount of attention to giving augmented reality (AR) glasses to kids in classrooms. The glasses act as a screen that projects information or images on them, supplementing, say, a book or map. Even better, AR glasses could provide a video of the teacher giving instructions to each kid individually.

TEACHING TECHNOLOGY TRAPS

That said, while some pretty neat technological developments are happening in education, increasingly, there are also concerns about what technology in the classroom might *take away* from education. There's the fear that technology in the classroom has had no real impact on student achievement or test scores.

There's also trepidation about forcing districts and schools to upgrade technology prematurely with little idea of how it's best used, and whether its implementation is effective for spurring learning or comes at the expense of basic reading, writing, and math skills. Even worse, those upgrades can leave less money for teachers' salaries and raises.

With that comes the worry that teachers will become

obsolete and the classroom will become an entirely automated experience. Already there is a concern that dependence on technology allows students to have a lot more information at hand, but also to collect it too easily. Having a broad, easily accessible wealth of knowledge might harm the deeper understanding that comes from a careful, analytical research process.

Regardless of whether these concerns are well-founded or not, there's no doubt technology will play a large and growing part in the education system.

FUN IN THE FUTURE:
WHAT ENTERTAINMENT WILL
LOOK LIKE IN 2050

Reality could completely change between now and 2050; it could become "augmented." The idea of augmented reality (AR) has been around since at least the 1960s, when researcher Ivan Sutherland—better known as the father of computer graphics—envisioned in his paper "The Ultimate Display" that a blending of digital information and human vision would create the illusion of being able to peer through walls. By the early twenty-first century, Columbia University research-ers had developed a bulky but wearable satellite-dish-equipped rig that enabled a user to peer through special sunglasses and see pop-up graphics about places in a New York neighborhood.

Since then, AR projects and applications have popped up everywhere. For example, the Pentagon's DARPA has been working on AR-enabled contact lenses. Such devices will be able to read digital information embedded in the landscape itself in the form of radio-frequency identification (RFID) tags attached to objects, buildings, and even people.

That's all nice, but it doesn't sound all that fun. For a moment, let's leave the more serious predictions to others and examine how we will entertain ourselves in the future. Will we still like to tilt back a cold one at our neighborhood pubs? Will we still be shaking what our mommas gave us in dance clubs? Or might we square off against robots in sports games?

HOLY HOLOGRAM! WE WILL BE VIRTUALLY INSIDE OUR ENTERTAINMENT

Your flat-screen TV may represent the epitome of entertainment right now, but by 2050, it will seem hopelessly outdated. In 2050, we'll likely demand to interact with our entertainment via virtual reality, instead of having it contained on the screen. Imagine playing a video game about World War II: you and your friends will have the option of hopping off the couch and storming the beaches of Normandy with everybody else. With this same technology, your children will be able to interact with their favorite fuzzy friends on TV by inviting them into the living room to dance around.

Entertainment won't be the only way we'll use virtual technology, however. Likely, we'll be able to meet up with friends and family around the world thanks to hologram technology. Let's say you have a business meeting with colleagues from New York, Seattle, and Beijing. All of you can meet up in one office to discuss the matter at hand. Long-distance relationships will become a little more manageable because of virtual visits, and all of your pals can show up to your destination wedding, no matter where they live.

OUR WORLD, BUT BETTER

With pervasive Wi-Fi Internet poised to take over the United States and the world, powering the next generations past 4G connectivity, the truly powerful simulated world will one day soon be no different from the world that we already live in, only better. It will be more informed, more personalized, and, above all, more empowering to us as individuals and as a civilization to put that information to its best use. And all of this information and dynamic possibility will be brought to you by the performance power of the supercomputers we're only now bringing to life. We'll explore this more in later sections.

THE PEN IS NO LONGER MIGHTIER: WE WILL RETURN TO ORAL STORYTELLING

Currently, many of us spend our spare time using social media like Facebook, Twitter, and blogs, media that allow us to type a quick message to let our friends know what we're doing. However, writing and reading those messages may not be how we communicate with our pals in the future. In fact, we may not read or write at all, a future that many dying newspapers are already confronting.

Futurist William Crossman believes that spoken language will replace written communication in the coming years, meaning that we won't need to teach children how to read and write, but rather how to use computers and think creatively. Crossman envisions a world in which we all use voice-in, voice-out (VIVO) computers. Everything we need to communicate will be handled by these machines. Rather than writing our memoirs, for example, we'd sit down in front of a webcam and tell our stories. As the recent rise of reality television and YouTube superstars bears out, many people will happily sit and watch a complete stranger spin a tale.

THE FUTURE OF VOICE MAIL

In some ways, voice mail has always been the precursor to VIVO, but over the phone. Might voice mail one day read a caller's emotions to give the message recipient a heads-up? The answer is yes, and such future voice mail technical advances are already in development. Some emerging technologies are focused on recognizing emotions in the voices of those who leave messages on their voice mail systems. One system, called Emotive Alert, has been under development at the Massachusetts Institute of Technology for several years.

The system allows users to decide which messages are the most urgent, helping them cut through the glut of electronic communications and be more efficient. It could learn to recognize basic moods—such as urgent, happy, formal, or excited—by analyzing variables such as speech rate and volume, comparing those variables with a database, and then spitting out an opinion. As communication systems become more integrated, it's a sure thing that voice mail will remain part of the mix. After all, it deals in that most primal type of human communication—speech—by digitizing it and feeding it into the electronic world we live and work in today. Voice mail, one of the first high-tech messaging systems, is here to stay.

Crossman's future means that vast swaths of the population will be illiterate, but should those people need to read something, their computer could scan and read it to them. In some ways, this may make communication more democratic, and with the predicted explosion in populations, it may even be necessary. Computers would have instantaneous translation services, making it easier for an urban resident to connect with a rural resident a world away.

IIII ANDROID ATHLETICS: FACING OFF AGAINST ROBOTIC RIVALS

It's difficult to even comprehend all the ways we might interact with robots by 2050. For example, robots may be conducting routine surgeries or piloting our airplanes. They could be conducting search-and-rescue missions or fighting in wars. One researcher even predicts that by 2050, we could be having sex with and marrying robots.

But when it comes to having fun either now or in 2050, you'd be hard-pressed to find a better way to spend a Saturday or Sunday than outside in the beautiful sun enjoying a sporting event. What might change by 2050, though, is who's playing. Roboticists predict that by 2050, they will have developed autonomous robots that will be able to beat the best soccer players in the world. That's right: by 2050, we could see robots versus humans in the contest for the World Cup. Now that would fill some bars and basements!

These robots won't just be wired with the steps for winning a soccer game. Rather, roboticists are working now to train robots how to play soccer by using human models.

The robots are presented with data that shows how humans respond to a series of soccer plays, so that when the robots are presented with the same scenarios, they have choices in how to respond. These robots will be able to perceive the play and act accordingly. If you want a glimpse of the future, check out the RoboCup games, in which researchers test out their current "players" and share information with one another. So far, there are no robotic concession workers for these events.

DINING ON DELICACIES: JELLYFISH RESTAURANTS AND OTHER EXOTIC EATS IN THE FUTURE

It's hard to imagine a time when friends and family won't meet up for a bite to eat, so of course we're including dining out on our list. However, don't plan on heading to a local steakhouse or burger joint. Because of factors including land use, population, and water supply, the American diet will undergo some changes by 2050. We'll be eating a lot more grains and beans and a lot less meat and dairy. We'll still eat our veggies, of course, but we won't have the wide array to choose from that we enjoy today. Futurists estimate that about 15 percent of our diet will come from animal products, and the rest will come from plants; the United States as a whole may have to stop exporting food by 2025.

If a juicy burger isn't an option, where will those animal

products come from? One possibility is that we'll eat more fish, but we certainly won't be eating the tuna and cod that we're accustomed to. According to reports published in *Science* in 2006, commercial fish stocks could be completely decimated by 2050 because of overfishing and pollution. Our fishing throws aquatic ecosystems off balance in irreparable ways. For example, by fishing for tuna, marlin, and swordfish, we eliminate the major predators of jellyfish.

As a result, jellyfish numbers are increasing rapidly, which may mean that we'll have to develop a taste for jellyfish sushi. Jellyfish have been eaten for thousands of years in Asian countries, but the taste may require some getting used to—one biology professor described it as like "tough strips of cucumber." One slightly more palatable option we might have in 2050 is squid, which unfortunately is very high in cholesterol.

Speaking of high cholesterol, let's check out how our health will look in the future. Unfortunately, that topic is not all fun and games.

HAPPY AND HEALTHY? HOW MEDICINE AND OUR HEALTH WILL CHANGE

In 2009, the United States spent about $7,960 per person on health care. By contrast, many of the other countries in the Organization for Economic Co-operation and Development (OECD) spent $3,233 per person. In 2013, U.S. healthcare spending per capita rose to $8,233, yet spending more money hasn't made things better. And we have growing rates of obesity, heart disease, type 2 diabetes, and other diseases that are caused or influenced by poor health habits. The vast majority of Americans agree that our healthcare system is on the brink of collapse. Something has to change, but what? Here are five predictions futurists are making about healthcare.

DOPING UP: DRUGS MAY ENHANCE OUR MENTAL ABILITIES

Media outlets frequently explore the issue of growing numbers of college students abusing off-label drugs—like Adderall, which is used to treat attention deficit hyperactivity disorder, and Provigil, which is used to treat narcolepsy—in hopes of boosting their focus and ability to cram for exams. And students are not the only ones.

In an informal 2008 survey in *Nature*, about one in five scientists who responded admitted to experimenting with nootropics, a controversial class of drugs that are believed to boost brain performance by altering the availability of neurochemicals, increasing brain cell metabolism, improving oxygen supply to the brain, or stimulating the growth of neurons.

As "brain doping" becomes more common, new nootropic chemicals—some available without a prescription—are emerging. One such product, Onnit Labs's Alpha BRAIN, contains ingredients that supposedly boost the brain's levels of the naturally occurring neurotransmitter acetylcholine.

The *Atlantic* writer Ari LeVaux, who experimented with taking Alpha BRAIN, reported that after taking the substance, he had unusually vivid dreams and awoke earlier than usual the next morning feeling more refreshed and alert. LeVaux also noticed that he was "slightly more organized" and had "a curious sense of emotional stability." While the experts that LeVaux consulted told him that there wasn't any evidence that occasional use of nootropics carried any risks, there isn't much information yet available about the longer-term effects of habitual use.

||||| GET WELL AND STAY WELL

One of the issues with health care today, some futurists reason, is that it focuses more on reactive medicine. When you get hurt or sick, you go to the doctor and get treatment. That's fine for some things, but if that random pain turns out to be a serious condition, you (and your insurance company) could be paying out a lot of money. That's why health care will ultimately shift its focus to preventive instead of reactive care. This means keeping you from getting sick in the first place, as well as catching things early before they become big problems.

Obviously, not everything can be prevented, but you can lessen the risk factors for the most common diseases. Heart disease, for example, is the leading cause of death in the United States. Some of the risk factors include smoking, high cholesterol, high blood pressure, obesity, and inactivity—and all of them are actionable. But many people don't realize the risks until they have a heart attack.

We often ignore warning signs, and we don't always get regular checkups (sometimes because we can't afford them). We also aren't very good about educating ourselves. Not only is this bad for us personally, but it also puts a strain on the healthcare system—treating an illness costs much more in both time and money than taking steps to prevent it in the first place. Futurists like Jim Carroll believe that within the next ten years or so, wellness programs will be considered an integral part of most organizations. They'll even offer incentives for employees who participate and meet their goals, and penalize those who don't.

IIIIᵢI CHECK YOUR OWNER'S MANUAL

Some diseases and conditions are hereditary. If you know your family's health history, you know that you have a higher risk of developing them. But even with that knowledge, you could be at risk for something and not be aware of it. One way to find out is to have your genes mapped. The Human Genome Project, completed in 2003, mapped the twenty to twenty-five thousand genes that make up our DNA.

This includes all the variations: each of us has a unique genetic sequence that can reveal whether we have the gene known to cause a certain condition or whether that gene has mutated. If you find out early enough that you carry the gene for a certain type of cancer, for example, you would know what to look for if it develops. Or if you're thinking about having a child, you'll already know whether there's a chance of passing something on.

As of this writing, getting your individual genetic code mapped costs around $3,000 and takes about a week. That's not exactly accessible for most people, and insurance doesn't cover it. However, advances in technology are driving down both the length of time it takes and the cost. The current goal is $1,000, which is about the cost of an MRI today.

But futurist Michio Kaku believes that soon it will only cost about $100 to map your genes and place them on a CD, making the process a part of routine medical practice. Your doctor will take a saliva or blood sample and almost instantaneously have a sort of owner's manual for your body. Known as genomic medicine, this practice would allow doctors to both treat illnesses based on your genes and work to prevent you from getting sick in the first place.

TAKING CHARGE: CONSUMER-DRIVEN HEALTH

Futurists predict that technology will soon be a major player in how we take charge of our own health. Sure, there are apps available now to help you monitor your weight, food intake, and other health conditions. But futurist Jim Carroll foresees technology functioning as a "dashboard for the human body." Not only will we be using phone apps, we'll also be integrating lots of gadgets to help monitor our health.

Technologies, such as scales that measure weight and body fat and send them via Wi-Fi to a website or an armband that measures blood pressure and plugs into an iPhone, will be more common. We'll be able to send this information to doctors so they can monitor our health all the time.

CAN AN ARTIFICIAL INTELLIGENCE SYSTEM DIAGNOSE YOUR SYMPTOMS BETTER THAN A HUMAN CAN?

In 2011, an IBM supercomputer called Watson demonstrated its artificial intelligence prowess by handily defeating two flesh-and-blood champions on the TV game show *Jeopardy*. But Watson has the potential to do a lot more than win trivia contests. Its developers envision using the system's ability to store and retrieve data to amass what could become the ultimate digital collection of medical data in existence.

Unlike less powerful systems, Watson has the ability to answer questions and analyze information in natural language—that is, the way that humans express themselves—and generate all the possible diagnoses that might come from the information. Not only that, but Watson actually can rank the diagnoses according to its understanding of the medical knowledge in textbooks, medical journals, and medical case reports.

In one demonstration of Watson's diagnostic abilities, researchers gave the system a fictitious case involving a patient who lived in Connecticut and had blurred vision and a family history of arthritis. Watson responded with a list of possible causes, topped by

Lyme disease—a seemingly unlikely diagnosis, but the one that researchers were looking for. You'll read more on these robot doctors in the next section.

Even further into the future, Michio Kaku believes that disease-detecting silicon chips could be encoded with DNA and placed in your bathroom mirror: "You blow on the bathroom mirror. It analyzes your saliva droplets, looking for the P53 gene." A change in this gene is present in half of the most common cancers. Frank Moss, former director of the MIT Media Lab, sees a future in which software could suggest how you might improve your health, based on your vital signs and other input.

A program might direct you to run diagnostic tests on yourself and then send the results to your doctor, who would explain them and prescribe treatment via video. And if you do have to go into a doctor's office, he or she will show you what's happening on a big screen. You'll decide on a course of action together, and you'll get instructions sent to your phone. The future of health technology will mean an entirely new doctor-patient relationship.

⦀ IF I ONLY HAD A (NEW) BRAIN: GROWING NEW ORGANS

Regenerative medicine involves healing the body by replacing or regenerating cells, tissues, or organs. It includes things like stem-cell and bone-marrow transplants, as well as artificial organs and medical devices. But there are currently some limits to this field. Artificial hearts are only used when a patient is about to die because they simply don't last long enough. If you need a new organ that can be transplanted, like a kidney, you may be on dialysis for years, at a cost of up to $30,000 per year. About eighteen people die each day (6,570 each year) while waiting for an organ transplant. If you do find a donor and get a new organ, you may have to contend with issues such as rejection.

But what if instead of getting a donated organ or having an artificial medical device implanted, you could just grow a new one? Futurists believe that this could be a possibility in ten to twenty years. Growing organs will not only save lives, it will also save the cost of treatments. Our aging population will require better and more efficient ways of managing their health, and a sort of "body shop" where you can replace what wears out will contribute to that.

This is probably not as far away as you might think. In 2008, British doctors implanted the first engineered organ, a new windpipe that was grown from the patient's own stem cells. We're already using tissue-engineered skin for burn patients, and there are clinical trials growing all kinds of other human cells, organs, and tissues—from ears to blood vessels.

IIII PURGING PAPER WORK: COORDINATED HEALTH CARE

If you've ever been admitted to a hospital, you know that for all of our automation, there's still a lot of actual paper work involved. You may also see lots of different healthcare providers during your stay, and sometimes they may give you conflicting information or not keep appointments. This is just one of the many aspects of a system that is inefficient and ultimately costs you more. Futurist Jack Uldrich predicts that hospitals will become more accountable and efficient by providing coordinated, standardized care.

Some of this has to do with technology. Healthcare providers need to get on board with digitizing everything from patient records to MRIs and electronically tracking things like pharmacy inventory and even people (in the case of, say, an Alzheimer's patient who tends to wander). Digitization will also let hospitals easily monitor things like outbreaks of illnesses.

Providing more efficient care also involves working together—the Patient Protection and Affordable Care Act

(PPACA), for example, gives hospitals incentives to form accountable care organizations. These are networks that include doctors, hospitals, and clinics that share accountability for the patients' care.

COULD TEXT MESSAGING REVOLUTIONIZE HEALTH CARE?

Many people in the healthcare industry are looking to use text messages as a new health tool to spread and communicate important health information, especially to communities that are traditionally underserved by medical professionals, such as rural populations, low-income areas, or areas with large minority populations.

In 2011, the U.S. Department of Health and Human Services helped support the launch of a program called Text4baby, which sends free health information and links to resources to mothers via text during their pregnancies and through their baby's first year of life. Researchers at the University of California, San Diego, and California State University, San Marcos, evaluated the program in November 2011 and found promising results. Most moms who signed up said that they learned valuable information, appreciated the reminders to make appointments and immunizations, and talked to their doctors about topics sent to them.

Many doctors are also using text messages to answer patient questions, give information on how to manage a specific condition, and remind patients of appointments or when to take medication. But again, it's important that physicians safeguard any private information sent via text just as they would in a phone call, or they could face lawsuits.

Certain companies, such as ER Texting, have also launched texting programs that can help cut down on your wait time in the ER and allow victims of natural disasters to communicate quickly with relief forces to get crucial medical assistance by texting a special code for free.

Joe Flower, another healthcare futurist, believes that hospitals will also improve by adopting management tools used in successful businesses, like Six Sigma or the Toyota Production System (yes, the car manufacturer). Some hospitals are already employing this model and report that they're better able to care for a larger number of patients without more staff or larger facilities.

TWEETS FOR HEALTH

Sure, you can send a mass email or a group text to your friends and family. But with Twitter, you don't have to input all those names or addresses. Instead, everyone who subscribes to your account sees your update. If your profile is public, everyone in the world can see your tweets. Just as with an email, your Twitter followers can chime in with a response, which is also public. Perhaps the power of Twitter will be harnessed by the medical community for good as well.

DIGITAL DOCTORS: WILL COMPUTERS REPLACE DOCTORS?

In real life, we already consult computers instead of doctors, using the Internet to research symptoms and conditions before we even think about making an appointment. That's still not the same as trusting an actual person, though, and some people argue that computers can never replace doctors for that reason. Computers can't have long conversations with patients to elicit bits of information that may be helpful in forming a diagnosis, nor can they show empathy to make that patient feel comfortable. No, we don't go to the doctor to make friends, but we don't want to feel like a number, either.

Will medicine as a profession be totally obsolete in the future? Probably not. Will computers replace doctors for some things? Probably so. The only thing that's certain is that medicine as we know it will change drastically in the future. Let's check out what that may look like.

MAN VERSUS MACHINE

There's no doubt that computers are becoming more

sophisticated, and plenty of people believe that they'll replace doctors to some extent. Futurist Kent Bottles, MD, argues that "within five years, primary-care providers will be replaced by sociable humanoid robots, avatars, and computer programs." He believes there will still be doctors, but they'll all specialize as computers take over the basic functions of a primary doctor.

Many diseases—what Bottles calls "rules-based chronic diseases" because we can understand them very well from a scientific standpoint—can be diagnosed and treated by a computer, including type 2 diabetes. This leaves doctors free to diagnose and treat chronic diseases that require more intuition and skill, like schizophrenia.

Farhad Manjoo, a technical writer whose wife is a pathologist, agrees that computers will replace doctors, but it's not the primary-care providers who should be concerned. Doctors who are specialists—like his wife—are the ones whose jobs are in trouble. Manjoo reasons that "robots are great specialists. They excel at doing one thing repeatedly, and when they focus, they can achieve near perfection." Because primary doctors treat a wide variety of diseases and conditions, they're versatile enough to keep their jobs. They also have those all-important communication skills.

Or do they? Journalist Ezra Klein points out that doctors, while good at the science part of practicing medicine, aren't always good at the conversation part. Computers can separate "conversation, the human element of the practice, from the technical diagnosis." Klein says that doctors' offices already have people who are great at the former: nurses. So maybe some primary-care physicians will lose their jobs, but we'll still need some kind of human healthcare professional to talk to us and feed our information into a computer. The medical profession will thus become more like other professions, such as accountancy, that have traditionally combined computer and human elements.

In fact, computers are already doing a lot of work in medicine. Here's a look at some examples.

AI ASSISTANCE: MEDICAL ROBOTS

Since the first robot-assisted surgery in 1985, robotic surgery has gotten a lot of attention. Today, the most common system is the da Vinci Surgical System. It has a 3-D camera system that surgeons view on a screen, as well as robotic arms that hold instruments inside the patient's body. The system translates a surgeon's hand movements, using robotic arms, into much smaller movements inside, allowing for tiny incisions in surgeries that would otherwise require large ones (and much more discomfort and recovery time).

This system is used in more than eight hundred hospitals in the United States and Europe. But until 2010, none of these robotic surgeries was solely performed by robots—it was always a robot mirroring the movements of a surgeon. Some robotic surgeries in 2010, however, were "fully robot-ic"—a surgeon pressed some buttons but did not manipulate the instruments in any way.

Robots are also used to help diagnose and treat disease. The Pap smear is a test that most women are familiar with. It helps diagnose cervical cancer, and women may get one every three years, starting at age twenty-one. The slides are often analyzed by an automated system, which targets areas of concern for a person to review later. This process has been proven to help catch more instances of precancerous lesions, leading to more computer-aided disease diagnostics.

Watson, the artificial intelligence IBM computer famous for beating *Jeopardy* contestants in 2011, is also being used to help diagnose conditions. The machine can use its evidence-based learning and natural language capabilities to receive a query from a doctor and mine the latest medical data to help reach a diagnosis.

But the question remains: would people ever accept just speaking to a computer instead of a doctor? Bottles thinks so. He cites a medical kiosk avatar featured at a panel called "Man-Made Minds: Living with Thinking Machines" at the World Science Festival in 2011. The kiosk was used by a young mother who was concerned about her child's diarrhea. The panel moderator stated that "the avatar was much more compassionate in relating to the child and his mother than human triage nurses she has encountered as a mother taking her child to New York City hospital emergency rooms."

As long as we can feel cared about and we're getting accurate diagnoses and treatment, maybe it doesn't matter whether we're receiving medical care from a human or a robot.

MIND OVER MATTER: TELEKINESIS, MIND CONTROL, AND DIGITAL IMMORTALITY

Yep, that's right: in our increasingly digital world, we could one day move objects with our minds and read the minds of others. In fact, not all that far in the future, our minds may be alive forever! Is your mind sufficiently blown yet?

If all of this seems too outrageous to comprehend, we concur. But it is happening—and faster than we think. Let's break it down into more manageable bytes (pun intended) and look at the developing technology behind each of these wild ideas. First, let's start with an age-old concept: telekinesis, or psychokinesis.

You've probably heard of so-called spoon-bending psychics who claim to possess psychokinesis—that is, the power to manipulate inanimate objects with their thoughts. While those folks may not actually possess such powers, in recent years scientists have made breakthroughs that someday may give all of us the ability to operate machines not by flipping a switch or manipulating a joystick, but by simply thinking about them!

The key to such power is something called a brain machine interface (BMI), which essentially is a communication pathway that allows your neurons to send signals to external gadgetry, just as easily as they do to your muscles. Starting in the 1970s and 1980s, researchers began developing algorithms, or mathematical formulas, that mimicked the brain's control over the muscles.

By the mid-2000s, they had begun to devise electronic brain implants called neuroprostheses, which picked up and translated human neural impulses into signals that could tell a robotic arm to move or manipulate a cursor on a computer screen. The technology is still in its infancy, but scientists envision someday equipping paralyzed people with neuroprostheses that will enable them to control powered exoskeletons to walk and do other everyday activities that fully abled people take for granted.

But others envision that someday, not only will we be able to turn the stove off or start the car by thinking about it, we'll be wirelessly connected to thought-controlled computers and devices that will continually provide us with information—for example, the names of people whose faces we can't place.

▐▐▐▌ MAPPING OUR MINDS: ULTRASONIC MIND CONTROL

Moving objects simply by thinking about them is fun, but how about mapping your own mind or someone else's, or perhaps controlling and changing it? The idea of anyone messing with your mind probably makes you nervous. But what if doctors could put that power to good use without drilling a hole through your skull?

"It's all in your head." You've heard that before, haven't you? Maybe someone was scoffing at a ghost story or downplaying the symptoms of depression, paranoia, or madness. The message is simple: your obsession, whatever it may be, has no basis outside your thoughts.

Yet with sufficient scientific understanding, it becomes obvious that everything is "all in your head." We don't mean that the world is one great big illusion, but rather that each individual's pool of consciousness and memory exists solely within the electrochemical processes of the brain. Self is a chemical cocktail imbued with a neurological spark, and consciousness is a peculiar parlor trick stirred up by evolution.

That person you think you are? Well, it's the product of a constantly changing equation made up of ninety-five to one hundred billion neurons, along with synapses, neurotransmitters, genetic coding, and a string of memories tailing back to the murky depths of childhood. Alter any factor in this equation just a little, and you change the final sum.

In fact, you do a little self-tinkering every time you so much as make a simple observation or drink a cup of coffee. Each successive you is at least a slight variant on the previous incarnation. The mind is also subject to the severe alterations of emotional trauma, brain injury, and disease, all capable of drastically changing the outcome of the neurological equation.

Our tools for addressing brain conditions have ranged from the sublime to the barbaric. We've treated mental illness on the therapist's couch, as well as with scalpels and electric shock. What if there were a way to stimulate nerves and portions of the human brain without drilling through skulls and implanting electrical devices? What if there were a way to tinker remotely with the neurological equation? Fortunately, scientific breakthroughs continue to refine our methods.

STIMULATING THE PSYCHE

There are several ways to stimulate your brain. There's neurostimulation, or electrically stimulating nerves to relieve pain or suppress tremors. Doctors accomplish these feats with the aid of tiny neurostimulators implanted near the spinal cord or a major nerve. Deep brain stimulation and vagus nerve stimulation take this concept even further and may effectively manage various psychiatric disorders and neurological diseases. The only catch is they require painstaking surgery to position implants in the appropriate locations—in the neck for vagus nerve stimulation and in the brain for deep brain stimulation.

A WINDOW INTO YOUR WORLD: SEEING INSIDE THE HUMAN MIND WITH ULTRASOUND

If you've ever experienced an ultrasound or studied the fetal images the technique can produce, you may have witnessed the future of neurostimulation.

Ultrasound operates like submarine sonar systems or bats, both of which emit sound waves to sense their surroundings. The waves travel until they make contact with an object, then bounce back to the source. A bat or a computer can then determine the shape and distance of the object based on this returning sound wave.

Ultrasound imaging systems transmit high-frequency sound pulses through the human body. Every time they hit a boundary between tissues, some bounce back while others keep going. The machine then calculates the distances and frequencies involved and creates a 2-D image of what's going on inside the body cavity—such as the movements of a fetus in utero.

Scientists have studied the effects of ultrasound on biological tissues since the 1920s. As early as the 1950s, researchers realized that at sufficiently high frequencies (much higher than those used in prenatal care), ultrasound also had the potential to destroy specific cells, especially tumors in the brain. Here's some more recent history:

- **High-frequency ultrasound** (HIFU) promised all of this without harming surrounding tissue or drilling a hole through a patient's skull. For decades, however, researchers lacked sufficient imaging technology to see what was specifically happening in the brain.

- Modern researchers, however, have advanced **magnetic resonance imaging** (MRI) to document the real-time interworking of the human body. Furthermore, brain-mapping technology continues to illuminate what's going on in the human mind. To return to the equation analogy, this means knowing exactly which parts of the neural equation affect which aspects of our abilities, memory, and personality.

- In a 2008 study by neuroscientists at Arizona State University, researchers discovered that **low-intensity, low-frequency ultrasound** (LILFU) could apply a gentler touch. Instead of destroying cells, these lower frequencies merely stimulate brain circuit activity.

IIII| LOW-FREQUENCY ULTRASOUND: THERAPY AND WEAPON OF THE FUTURE?

To understand how LILFU stimulates brain circuit activity, you have to grasp what's going on with your gray matter. Brain cells release neurotransmitters, molecules that carry information from one nerve cell to another across small gaps called synapses. When they arrive at another cell, neurotransmitters cause ion channels to open, which in turn triggers the electrical impulses that pass messages along nerve fibers. These reactions are a vital component of the brain's circuitry, and neurotransmitter disruptions are symptoms of such debilitating conditions as Alzheimer's disease, Parkinson's disease, depression, and epilepsy.

The team of neuroscientists at Arizona State University found that LILFU waves boosted the release of neurotransmitters, possibly by opening up sodium and calcium ion channels enough to trigger action potentials, which in turn release neurotransmitters. This means that, without invasive surgery, physicians in the future may be able to undo the damage produced by diseases such as Alzheimer's by stimulating the production of the very neurotransmitters that the condition disrupts.

It will take years of research and development before LILFU technology is ready to alter human brain circuitry and nonsurgically repair neurological injuries and diseases.

As you might imagine, however, the effects of ultra-sound on the human brain haven't inspired only therapeutic innovations. Various studies have reportedly theorized that ultrasonic weapons could be used to induce vomiting and nausea in victims, perhaps during a riot-control scenario. Other scientists suggest that the technology could eventually allow us to manipulate human memory.

Ultrasonic mind control is a science very much in its infancy. With enough research, we might one day live in an age where police use ultrasound to incapacitate you at a protest, while physicians optimize human brain efficiency with a little ultrasonic fine-tuning.

THE FOUNTAIN OF YOUTH: DIGITAL IMMORTALITY

Perhaps the biggest limitation of the human intellect is its shelf life. You only have so many years to learn, because no matter how smart you get, eventually the body that carries around your observations and insights will die. Humans have tried to overcome that by writing books and amassing libraries to pass along knowledge, but it's hard to preserve more than a smidgen of the data stored in the roughly one hundred billion neurons in the typical human brain.

But some futurists see a way around that. What if we could capture and digitize all the information in our brains and then upload that data to a computer or a robot?

HUMAN-TO-ROBOT TRANSPLANT

Russian industrialist and media mogul Dmitry Itskov hopes to achieve a cruder work-around version of that vision—transplanting a working human brain into a robot—in just a decade through his 2045 Initiative. But that's just the first step. Within thirty years, Itskov envisions finding a method of copying and uploading human consciousness into a machine, or even a holographic virtual body—basically, a software replica of a person.

That may sound totally, impossibly crazy. But given researchers' recent progress in developing neurosynaptic computer chips—that is, machines that mimic the neurons and synapses of the brain—it's hard to just scoff at such a bold prediction. Such chips eventually may have the ability not just to store information, but also to learn and remember, just as real brain cells do. That could mean that we'll not only be able to create complete copies of our brains' content, but that those copies would be able to keep using what we know and build upon it, long after our original bodies have vanished.

Imagine, for example, how many more great plays William Shakespeare might have written if he'd had another century to further develop his craft and to find new sources of inspiration. Or better yet, how many new dramas he'd be producing today if a copy of his brain was still at work.

COMPUTING OUR CULTURE: SUPERCOMPUTING TO MAKE US SUPERHUMANS

In order to revolutionize education, participate in our virtual video games, heal our broken healthcare system, or become digital immortals, we first need supercomputers that can support all these developments. Let's take a look at those now, starting with the first supercomputer: the CDC 6600.

Most of us are familiar with Moore's law, which states that the number of transistors on a 1-inch (2.5 centimeter) diameter of silicon doubles every twenty-four months. It's easy to forget that this doesn't just apply to our laptops and desktops. All of our electronic devices benefit from the same improvement cycle: processing speed, sensor sensitivity, memory, and even the pixels in your camera or phone. But chips can only get so small and powerful before certain effects due to quantum mechanics kick in, and some experts say this trend—which has rung surprisingly true over the last fifty years—may slow down over the next decade as we get closer to the actual limit of what our current materials can do.

Our phones and tablets may be the result of shrinking vast amounts of computing power down to something you can take to the beach, but we're only seeing the face of all that data. Behind the scenes, the "cloud" requires more and faster information and computation than ever before, with those needs rising just as steadily as the quality and amount of data we're enjoying on our side of the screen. From the high-def movies we stream to the weather, traffic, and other satellite info we already use daily, the future lies in plain old number crunching. And that's what super-computers do best.

IIII EXAFLOPS AND BEYOND!

Miniaturization of chip components is only half the story. On the other side, you have the supercomputer: custom setups, built for power. In 2008, the IBM Roadrunner broke the one-petaflop barrier: one quadrillion operations per second. (FLOPS stands for "floating-point operations per second," and it's the standard we use to talk about supercomputers used for scientific calculations, like the ones we're discussing here.)

Expressed in scientific notation, a petaflop is measured on a scale of 10^{15} operations per second. An exaflop computer—which experts predict will arrive by 2019—performs at 10^{18}, or a thousand times faster than the petaflop computers we're seeing now. Continuing into the future, zettaflops improve on the same scale, giving us 10^{21} operations per second by 2030, and then come the yottaflops, at 10^{24}.

But what do those numbers really mean? Well, for starters, it's believed that a complete simulation of the human brain will be possible by 2025, and within ten years, zettaflop computers should be able to accurately predict the entire weather system two weeks in advance.

TOP500.ORG

Since 1993, the TOP500 project has been the central resource for tracking and detecting supercomputing trends. Twice yearly, an ordered list of the most powerful computer systems currently operational is released. These rankings are based on a specific set of criteria, the Linpack benchmark, which involves running a supercomputer through several dense linear equations and measuring its response times. TOP500 appends each system's specs, as well as the areas in which it will be used.

At the 2012 International Supercomputing Conference in Hamburg, Germany, a U.S. supercomputer made the top of the list for the first time since 2009: IBM's Sequoia, which is installed at the Department of

Energy's Lawrence Livermore National Laboratory, came through with a Linpack score of 16.32 petaflops. Sequoia unseated Japan's K Computer, which had taken the top spot from China's Tianhe-1A the previous year, with a score in excess of 10 petaflops.

GREEN SUPERCOMPUTERS

All that power comes at a cost. If you've ever had a failed heat sink crash your desktop computer or sat with a laptop actually in your lap for more than a few minutes, you know what that cost is: computers use a lot of heat-producing energy. In fact, one of the major challenges for supercomputer developers is finding a sensible way to install and use the mighty machines without hardware failures or increased damage to the planet. After all, one of the main uses of weather simulations will be to monitor carbon and temperature fluctuations, so it wouldn't be very smart to add to the very problem climatologists are trying to solve!

Any computer system is only useful if it works, so keeping those hot circuits cool is of major interest. In fact, more than half of the energy used by supercomputers goes directly to cooling. But since the future of supercomputing is tied up with other vanguard sciences and futurist agendas, ecological concerns are already a matter of grave importance to high-performance computer engineers. Green solutions and energy efficiency are parts of every supercomputer project underway, and in fact, Sequoia's efficiency was a large secondary reason the IBM supercomputer debuted to such fanfare.

From cooling with "free air"—that is, engineering ways to get outside air into the system—to hardware designs that maximize surface area, scientists are trying to be as innovative with efficiency as they are with speed. One of the most interesting ideas, used by several teams, is to run the system in a conducting liquid that picks up the heat, then pipes the heat through the structure that houses the computer banks themselves. Heating water and rooms while simultaneously cooling the equipment is an idea that has many applications beyond just supercomputer sites, but the projects on TOP500's list are taking ideas like these very seriously.

Addressing ecological and efficiency concerns is not only a good idea for our planet, but also necessary to make the machines run. It's perhaps not the most glamorous application this research can promise the rest of us, though—and you were promised futuristic trends!

THE ARTIFICIAL BRAIN

As previously mentioned, it won't be all that long before supercomputers are able to map the human brain. A scientist at Syracuse University estimated in 1996 that our brains have a memory capacity somewhere between 1 and 10 terabytes, probably around 3. Of course, this wasn't a hugely useful comparison, since our brains don't work the same way computers do. But within the next twenty years, computers should be able to work the way our brains do!

In the same way supercomputers are already useful in mapping and affecting the human genome, producing solutions, and predicting inherited medical issues and predilections, accurate models of the human brain will mean huge leaps in diagnosis, treatment, and our understanding of the complexities of human thought and emotion.

In conjunction with imaging technology, doctors would be able to pinpoint trouble areas, simulate different forms of treatment, and even get to the root of many questions that have plagued us from the beginning of time. Chips that can be implanted and grafted and other technology could help monitor and even shift levels of serotonin and other brain chemicals related to mood disorders, while major brain malfunctions and injuries could simply be reversed.

Beyond the medical advances of this technology, there's also the matter of artificial intelligence (AI). Midperformance computing power already gives us some powerful AI. Think of the intelligent systems already recommending customized selections of television, movies, and books based on AI algorithms, or the hours we spend chatting with Siri and similar virtual-intelligence programs. But a humanlike level of "mental" complexity means applications for true AI.

Imagine a WebMD that actually responds like a doctor, bringing expert levels of attention to the questions you ask. Now expand that concept beyond medical concerns to imagine virtual experts explaining anything you need to know in a comfortable, conversational environment.

WILL TOUCH SCREENS EVER BE FINGERPRINT-PROOF?

The computers of the future certainly may be super, but the current ones in our smartphones and tablets come with some decidedly low-tech issues. Consider touch screens, which can be smudgy and hard to read or access.

Given our high-tech gadgets, this simple issue can be surprisingly annoying—and sometimes downright inhibiting. Although our fingers are useful for tapping and swiping at our screens, they're also dirty. The more we manipulate our gadgets or computers, the less pristine, sleek, and advanced they appear. With all the progressive electronics under the hood of our touch screens, will there ever be a way to create a touch-screen phone, tablet, or monitor free of the smudges, dirt, and grime we carry around on our hands?

The answer, in short, is…probably.

One solution is a self-cleaning paint. Titanium dioxide (used in sunscreens and cosmetics) is being studied for its ability to destroy microbes and germs on contact. Researchers believe that adding titanium dioxide to the surface of the touch screen could result in the reduced appearance of the oils that slough off your fingers.

Scientists in Germany have been developing a superamphiphobic surface, one that repels both water and oils. In a messy irony, they used soot to cover the surface they were testing and then blasted it with heat to harden it, using a chemical process called calcination. What resulted was a surface where oils and water rolled off into droplets, leaving no trace behind. The Japanese company Toray also has introduced a film that includes a wet coating where fingerprints won't stick, with an oil-repelling material for good measure.

Of course, there are ways to clean your screen in the meantime. You can certainly buy covers that will protect the screen from your fingers, but possibly at the risk of losing some of the clarity you see when you're using the screen. Follow directions about using soft cloths to wipe down your screen, and if you need something a bit stronger, spray the rag with distilled water or even a 50 percent mixture of water and vinegar.

Let's also acknowledge that Apple has already addressed this issue to some extent with an oleophobic (oil-repelling) touch screen. A patent filed in February 2011 is believed to improve on the method currently being used. We'll just have to wait and see if it gets the job done.

HOW WILL COMPUTERS EVOLVE OVER THE NEXT ONE HUNDRED YEARS?

To call the evolution of the computer meteoric seems like an understatement. Consider Moore's law, which, as we mentioned earlier, states that the number of transistors on a 1-inch (2.5 centimeter) diameter of silicon doubles every twenty-four months.

Awareness of the breakneck speed at which computer technology develops has seeped into the public consciousness. We've all heard the joke about buying a computer at the store only to find out it's obsolete by the time you get home. What will the future hold for computers?

Assuming microprocessor manufacturers can continue to live up to Moore's law, the processing power of our computers should double every two years. That would mean computers a hundred years from now would be 1,125,899,906,842,624 times more powerful than the current models. Hard to imagine!

But even Gordon Moore would caution against assuming Moore's law will hold out that long. In 2005, Moore said that as transistors reach the atomic scale, we may encounter fundamental barriers we can't cross. At that point, we won't be able to cram more transistors in the same amount of space.

We may get around that barrier by building larger processor chips with more transistors. But transistors generate heat, and a hot processor can cause a computer to shut down. Computers with fast processors need efficient cooling systems to avoid overheating. The larger the processor chip, the more heat the computer will generate when working at full speed.

Another tactic is to switch to multi-core architecture. A multi-core processor dedicates part of its processing power to each core. These processors are good at handling calculations that can be broken down into smaller components; however, they aren't as good at handling large computational problems that can't be broken down.

Future computers may rely on a completely different model than traditional machines. What if we simply abandon the old transistor-based processor?

OPTIMIZING OPTICS: FIBER OPTICS, QUANTUM PROCESSING, AND DNA COMPUTERS

Fiber-optic technology has already begun to revolutionize computers. Fiber-optic data lines carry information at incredible speeds and aren't vulnerable to electromagnetic interference like classic cables. What if we were to build a computer that uses light to transmit information instead of electricity?

One benefit is that an optical or photonic system would generate less heat than the traditional electronic transistor processor. The data would transmit at a faster rate as well. But engineers have yet to develop a compact optical transistor that can be mass-produced. Scientists at ETH Zurich were able to build an optical transistor just one molecule in size. But to make the system effective, the scientists had to cool the molecule to $-272°C$, or $1°K$. That's just a little warmer than deep space. And that's not really practical for the average computer user.

Photonic transistors could become part of a quantum computer. Unlike traditional computers, which use binary digits or bits to perform operations, quantum computers use quantum bits, or qubits. A bit is either a 0 or a 1. Think of it like a switch that is either off or on. But a qubit can be both a 0 and a 1 (or anything in between) at the same time. The switch is both off and on and everything in between.

A working quantum computer should be able to solve big problems that can be split into smaller ones much faster than a traditional computer. We call these problems embarrassingly parallel problems. But quantum computers are, by their very nature, unstable. If the quantum state of the computer is upset, the machine could revert to the computing power of a traditional computer. Like the optical transmitter created at ETH Zurich, quantum computers are kept at just a few degrees above absolute zero to preserve their quantum states.

Perhaps the future of computers lies inside of us. Teams of computer scientists are working to develop computers that use DNA to process information. This combination of computer science and biology could lead the way to the next generation of computers. A DNA computer might have several advantages over traditional machines. For example, DNA is a plentiful and cheap resource. If we discover a way to harness DNA as a data-processing tool, it could revolutionize the computer field.

THROUGH THE RABBIT HOLE

The world of quantum physics is a strange one, particularly if you're only familiar with classical physics. At the quantum level—which includes the world of atomic particles up to molecular structures—the normal rules don't apply. For example, we know that by observing a quantum event we change that event's outcome—so it's impossible to measure anything on the quantum scale without altering it in some way.

ⅢⅢ⌶ UBIQUITOUS COMPUTING

A popular theme in science-fiction stories set in the future is ubiquitous computing. In this future, computers have become so small and pervasive that they are in practically everything. You might have computer sensors in your floor that can monitor your physical health. Computers in your car can assist you when you drive to work. And computers practically everywhere track your every move.

It's a vision of the future that is both exhilarating and frightening. On the one hand, computer networks would become so robust that we'd always have a fast, reliable connection to the Internet. You could communicate with anyone you choose no matter where you are with no worries about interruption in service. But on the other hand, it would also become possible for corporations, governments, or other organizations to gather information about you and keep tabs on you wherever you go.

We've seen steps toward ubiquitous computing over the last decade. Municipal Wi-Fi projects and 4G technologies like LTE and WiMAX have extended network computing far beyond the world of wired machines. You can purchase a smartphone and access petabytes of information on the World Wide Web in a matter of seconds. Sensors in traffic stoplights and biometric devices can detect our presence. It may not be long before nearly everything we come into contact with has a computer or sensor inside it.

We may also see massive transformations in user interface technology. Currently, most computers rely on physical input interfaces like a computer mouse, keyboard, tracking pad, or other surface on which we input commands. There are also computer programs that can recognize your voice or track your eye movements to execute commands. Computer scientists and neurologists are working on various brain-computer interfaces that will allow people to manipulate computers using only their thoughts. Who knows? The computers of the future may react seamlessly with our desires.

To predict the next hundred years is difficult. Technological progress isn't necessarily linear or logarithmic. We may experience decades of progress followed by a period in which we make very little headway as we bump up against unforeseen barriers. On the other hand, according to some futurists, there may be no meaningful difference between computers and humans within a hundred years. In that world, we'll be transformed into a new species that can improve itself at a pace unimaginable to us in our current forms. Whatever the future may hold, it's safe to assume that the machines we rely on will be very different from today's computers.

HEADS IN THE CLOUDS: HOW CLOUD COMPUTING WORKS

You may have heard a lot of talk about clouds recently, and we're not referring to those angry gray masses of vapor that send torrents of rain down on our heads. We're talking about clouds for computers. Cloud computing allows you to access files and programs on your computer from across a network. It's really convenient, but what are you giving up in return?

These days, instead of installing a suite of software for each computer, individuals and corporations have the option of loading only one application. That application allows workers to log into a web-based service that hosts all the programs users would need for their jobs. Remote machines owned by another company run everything from email to word processing to complex data-analysis programs. It's called cloud computing, and it is changing the entire data technology industry.

In a cloud computing system, there's a significant work-load shift. Local computers no longer have to do all the heavy lifting when it comes to running applications. The network of computers that make up the cloud handles them instead. Hardware and software demands on the user's side decrease. The only thing the user's computer needs to be able to run is the cloud computing system's interface software, which can be as simple as a web browser, and the cloud's network takes care of the rest.

There's a good chance you've already used some form of cloud computing. If you have an account with a web-based email service like Gmail, Hotmail, or Yahoo! Mail, then you've had some experience with cloud computing. Instead of running an email program on your computer, you log in to a web email account remotely. The software and storage for your account don't exist on your computer—they're on the service's computer cloud.

So what makes up a cloud computing system? Let's find out!

THE ARCHITECTURE OF A CLOUD

When talking about a cloud computing system, it's helpful to divide it into two sections: the front end and the back end. They connect to each other through a network, usually the Internet. The front end is the side the computer user, or client, sees. The back end is the "cloud" section of the system.

The front end includes the client's computer (or computer network) and the application required to access the cloud computing system. Not all cloud computing systems have the same user interface. Services like web-based email programs leverage existing web browsers like Internet Explorer or Firefox. Other systems have unique applications that provide network access to clients.

On the back end of the system are the various computers, servers, and data storage systems that create the "cloud" of computing services. In theory, a cloud computing system could include practically any computer program you can imagine, from data processing to video games. Usually, each application will have its own dedicated server.

CLOUD COMPUTING: A CHEAT SHEET

Stuff you need to know:

][Cloud computing systems generally have a front end, which is what the user sees, and a back end, which does all the work.

][Cloud computing shares some similarities with an older model of computing called timesharing. A timesharing computer system connects multiple users to a single computer processor through dumb terminals, which have a keyboard and monitor, but leave the computing to the central machine.

][While cloud computing promises to off-load tasks

like data storage and processing power, the model raises questions about data accessibility and security. How can you ensure that you can get to your data and keep it safe if it's on someone else's computer?

A central server administers the system, monitoring traffic and client demands to ensure everything runs smoothly. It follows a set of rules called protocols and uses a special kind of software called middleware. Middleware allows networked computers to communicate with each other. Most of the time, servers don't run at full capacity. That means there's unused processing power going to waste. It's possible to fool a physical server into thinking it's actually multiple servers, each running with its own independent operating system. The technique is called server virtualization. By maximizing the output of individual servers, server virtualization reduces the need for more physical machines.

If a cloud computing company has a lot of clients, there's likely to be a high demand for a lot of storage space. Some companies require hundreds of digital storage devices. A cloud computing system needs at least twice the number of storage devices it requires to keep all its clients' information stored.

That's because these devices, like all computers, occasionally break down. A cloud computing system must make a copy of all its clients' information and store it on other devices. The copies enable the central server to access backup machines to retrieve data that otherwise would be unreachable. Making copies of data as a backup is called redundancy.

What are some of the applications of cloud computing? Read on to find out.

I COMPUTED LONELY AS A CLOUD

Although cloud computing is an emerging field of computer science, the idea has been around for a few years. It's called cloud computing because the data and applications exist on a "cloud" of web servers.

CLOUD COMPUTING APPLICATIONS

The applications of cloud computing are practically limitless. With the right middleware, a cloud computing system could execute all the programs a normal computer could run. Potentially, everything from generic word-processing software to customized computer programs designed for a specific company could work on a cloud computing system.

Why would anyone want to rely on another computer system to run programs and store data? Here are just a few reasons:

- **Clients are able to access their applications and data from anywhere at any time.** They can access the cloud computing system using any computer linked to the Internet. Data isn't confined to a hard drive on one user's computer or even a corporation's internal network.

- **It can bring hardware costs down.** Cloud computing systems reduce the need for advanced hardware on the client side. You no longer need to buy the fastest computer with the most memory because the cloud system takes care of those needs for you. Instead, you can buy an inexpensive computer terminal. The terminal could include a monitor, input devices like a keyboard and mouse, and just enough processing power to run the middleware necessary to connect to the cloud system. You also don't need a large hard drive because you'd store all your information on a remote computer.

- **Corporations that rely on computers have to make sure they have the right software in place to achieve goals.** Cloud computing systems give these organizations company-wide access to computer applications. The companies don't have to buy a set of software or software licenses for every employee. Instead, the company can pay a metered fee to a cloud computing company.

- **Servers and digital storage devices take up space.** Some companies rent physical space to store servers and databases because they don't have enough room available on site. Cloud computing gives these companies the option of storing data on someone else's hardware, removing the need for physical space on the front end.

- **Corporations might save money on IT support.** Streamlined hardware would, in theory, have fewer problems than a network of heterogeneous machines and operating systems.

- **If the cloud computing system's back end is a grid computing system, the client could take advantage of the entire network's processing power.** Often, scientists and researchers work with calculations so complex that it would take years for individual computers to complete them. On a grid computing system, the client could send the calculation to the cloud for processing. The cloud system would tap into the processing power of all available computers on the back end, significantly speeding up the calculation.

The cloud also can help you discover new information and entertainment: new books, new movies and TV shows, new podcasts and radio shows, and new music.

While the benefits of cloud computing seem convincing, are there any potential problems?

WHO'S WHO IN CLOUD COMPUTING

Some of the companies researching cloud computing are big names in the computer industry. Microsoft, IBM, and Google are investing millions of dollars into research. Some people think Apple might investigate the possibility of producing interface hardware for cloud computing systems.

CLOUD COMPUTING CONCERNS

Perhaps the biggest concerns about cloud computing are security and privacy. The idea of handing over important data to another company worries some people. Corporate executives might hesitate to take advantage of a cloud computing system because they can't keep their company's information under lock and key.

The counterargument is that the companies offering cloud computing services live and die by their reputations. It benefits these companies to have reliable security measures in place. Otherwise, the service would lose all its clients. It's in the cloud computing companies' interest to employ the most advanced techniques to protect their clients' data.

Privacy is another matter. If a client can log in from any location to access data and applications, it's possible the client's privacy could be compromised. Cloud computing companies will need to find better ways to protect and ensure client privacy. One way is to use authentication techniques such as usernames and passwords. Another is to employ an authorization format so each user can access only the data and applications relevant to his or her job.

Some questions regarding cloud computing are more philosophical. Does the user or company subscribing to the cloud computing service own the data? Does the cloud computing system, which provides the actual storage space, own it? Is it possible for a cloud computing company to deny a client access to that client's data? Several companies, law firms, and universities are debating these and other questions about the nature of cloud computing.

How will cloud computing affect other industries? There's a growing concern in the IT industry about how cloud computing could affect the business of computer maintenance and repair. If companies switch to using streamlined computer systems, they'll have fewer IT needs.

Some industry experts believe that the need for IT jobs will migrate to the back end of the cloud computing system.

Another area of research in the computer science community is autonomic computing. An autonomic computing system is self-managing, which means the system monitors itself and takes measures to prevent or repair problems. Currently, autonomic computing is mostly theoretical. But, if autonomic computing becomes a reality, it could eliminate the need for many IT maintenance jobs.

PRIVATE EYES ARE WATCHING YOU

There are a few standard hacker tricks that could cause cloud computing companies major headaches. One of those is called key logging. A key logging program records keystrokes. If a hacker manages to successfully load a key logging program onto a victim's computer, he or she can study the keystrokes to discover usernames and passwords. Of course, if the user's computer is just a streamlined terminal, it might be impossible to install the program in the first place.

2

POWER PLAYS:
COOL WAYS TO ENERGIZE
OUR FUTURE

Now that we've looked at some of the crazy, cool inventions of the future, let's dig into how we'll be able to power them and our lives as we move into the coming decades. Tons of energy ideas are being thrown around, but in this section, we'll focus on some of the biggest and most interesting ones: solar power, steam, gasification and biofuels, and hydrogen. Time to power up!

CHASING THE SUN: HOW SOLAR THERMAL POWER WORKS

Most of us don't think much about where our electricity comes from, only that it's available and plentiful. Electricity generated by burning fossil fuels such as coal, oil, and natural gas—the major sources of energy for us—emits carbon dioxide, nitrogen oxides, and sulfur oxides, gases scientists believe contribute to drastic climate change. Solar thermal (heat) energy is a carbon-free, renewable alternative to the power we generate with fossil fuels like coal and gas. This isn't something solely available in the future, either.

You may recall that an average megawatt is the amount of electricity produced over a period of one year. Between 1984 and 1991, the United States built nine solar energy plants in California's Mojave Desert, and today they continue to provide a combined capacity of 354 megawatts annually of clean, safe, and ultra-reliable power used in 500,000 Californian homes. In 2008, when six days of peak demand buckled the power grid and caused electricity outages in California, those solar thermal plants continued to produce at 110 percent capacity.

Wondering where the technology's been since then? In the 1990s, when prices of natural gas dropped, so did interest in solar thermal power. Today, though, the technology is poised for a comeback. It's estimated by the U.S. National Renewable Energy Laboratories that solar thermal power could provide hundreds of gigawatts of electricity annually, equal to more than 10 percent of demand in the United States.

Shake the image of solar panels from your head—that kind of demand is going to require power plants. Let's take a look at how solar thermal power can handle this task.

HARNESSING SUN POWER

There are two main ways of generating energy from the sun: photovoltaic (PV) and concentrating solar thermal (CST), also known as concentrating solar power (CSP) technologies.

PV converts sunlight directly into electricity. These solar cells are usually found powering devices such as watches, sunglasses, and backpacks, as well as providing power in remote areas.

Solar thermal technology is large scale by comparison. One big difference from PV is that solar thermal power plants generate electricity indirectly. Heat from the sun's rays is collected and used to heat a fluid. The steam produced from the heated fluid powers a generator that produces electricity. It's similar to the way fossil-fuel-burning power plants work, except the steam is produced by the collected heat rather than from the combustion of fossil fuels.

IIIIII HEAT RISES: HOW SOLAR THERMAL SYSTEMS WORK

There are two types of solar thermal systems: passive and active. A passive system requires no equipment, like when heat builds up inside your car when it's left parked in the sun. An active system requires some way to absorb and collect solar radiation and then store it.

Solar thermal power plants are active systems, and while there are a few types, they have some basic similarities. Mirrors reflect and concentrate sunlight, and receivers collect that solar energy and convert it into heat energy. A generator can then be used to produce electricity from this heat energy.

The most common type of solar thermal power plants, including those facilities in California's Mojave Desert, use a parabolic trough design to collect the sun's radiation. These collectors are known as linear concentrator systems, and the largest are able to generate 80 megawatts of electricity. They are shaped like a half-pipe you'd see used for snowboarding or skateboarding, and have linear, parabolic-shaped reflectors covered with more than 900,000 mirrors that are north-south aligned and able to pivot to follow the sun as it moves east to west during the day. Because of its shape, this type of plant can reach operating temperatures of about 750°F (400°C), concentrating the sun's rays at thirty to one hundred times their normal intensity onto heat-transfer-fluid or water/steam-filled pipes. The hot fluid is used to produce steam, and the steam then spins a turbine that powers a generator to make electricity. While parabolic trough designs can run at full power as solar energy plants, they're more often used as a solar and fossil-fuel hybrid, adding fossil-fuel capability as backup.

Solar power tower systems are another type of solar thermal system. Power towers rely on thousands of heliostats, which are large, flat sun-tracking mirrors, to focus and concentrate the sun's radiation onto a single tower-mounted receiver. Like parabolic troughs, heat-transfer fluid or water is heated in the receiver, eventually converted to steam, and used to produce electricity with a turbine and generator. Unlike parabolic troughs, power towers are able to concentrate the sun's energy as much as 1,500 times. Power tower designs are still in development, but they could one day be realized as grid-connected power plants producing about 200 megawatts of electricity per tower.

A third system is the solar dish/engine. Compared to the parabolic trough and power towers, dish systems are small producers (about 3 to 25 kilowatts). There are two main components: the solar concentrator (the dish) and the power conversion unit (the engine/generator). The dish is pointed at and tracks the sun and collects solar energy; it's able to concentrate that energy by about two thousand times. A thermal receiver, a series of tubes filled with a cooling fluid (such as hydrogen or helium), sits between the dish and the engine. It absorbs the concentrated solar energy from the dish, converts it to heat, and sends that heat to the engine where it becomes electricity.

FILL 'ER UP: SOLAR THERMAL HEAT STORAGE

Solar thermal systems are a promising renewable energy solution because the sun is an abundant resource—except when it's nighttime, or when the sun is blocked by cloud cover. Thermal energy storage (TES) systems are high-pressure liquid storage tanks used along with a solar thermal system to allow plants to bank several hours of potential electricity. Off-peak storage is a critical component to the effectiveness of solar thermal power plants.

Three primary TES technologies have been tested since the 1980s, when the first solar thermal power plants were constructed: a two-tank direct system, a two-tank indirect system, and a single-tank thermocline system.

- In a **two-tank direct system**, solar thermal energy is stored in the same heat-transfer fluid that collected it. The fluid is divided into two tanks, one tank storing it at a low temperature and the other at a high temperature. Fluid stored in the low-temperature tank runs through the power plant's solar collector, where it's reheated and sent to the high-temperature tank. Fluid stored at a high temperature is sent through a heat exchanger that produces steam, which is then used to produce electricity in the generator. And once it's been through the heat exchanger, the fluid returns to the low-temperature tank.

- A **two-tank indirect system** functions basically the same as the direct system except that it works with different types of heat-transfer fluids, usually those that are expensive or not intended for use as storage fluid. To overcome this, indirect systems pass low-temperature fluids through an additional heat exchanger.

- Unlike the two-tank systems, the **single-tank thermocline system** stores thermal energy as a solid, usually silica sand. Inside the single tank, parts of the solid are kept at low to high temperatures in a temperature gradient, depending on the flow of fluid. For storage purposes, hot heat-transfer fluid flows into the top of the tank and cools as it travels downward, exiting as a low-temperature liquid. To generate steam and produce electricity, the process is reversed.

Solar thermal systems that use mineral oil or molten salt as the heat-transfer medium are prime for TES, but unfortunately without further research, systems that run on water/steam aren't able to store thermal energy. Other advancements in heat-transfer fluids include research into alternative fluids, phase-change materials, and novel thermal storage concepts, all in an effort to reduce storage costs and improve performance and efficiency.

FLOWER POWER: SOLAR THERMAL GREENHOUSES

The idea of using thermal mass materials—materials that have the capacity to store heat—to store solar energy is applicable to more than just large-scale solar thermal power plants and storage facilities. The idea can work in something as commonplace as a greenhouse.

All greenhouses trap solar energy during the day, usually with the benefit of south-facing placement and a sloping roof to maximize sun exposure. But once the sun goes down, what's a grower to do? Solar thermal greenhouses are able to retain that thermal heat and use it to warm the greenhouse at night.

SOLAR POWER ON THE RISE

Solar cell manufacturers and suppliers believe PV technology will produce 15 percent of the energy the United States will consume in 2020. Solar power is growing in popularity around the world. In Japan, homes generated roughly 80 percent of the total 1.9 million kilowatts of solar energy produced in the fiscal year ending March 2008. Japan aims to increase its solar power output by 40 percent by 2030. Also by 2030, the U.S. National Center for Photovoltaics (NCPV) has set the goal of using solar energy to supply 10 percent of the nation's power during peak generating times, as well as supply solar energy to foreign markets.

Stones, cement, and water or water-filled barrels can all be used as simple, passive thermal mass materials (heat sinks), capturing the sun's heat during the day and radiating it back at night.

Bigger aspirations? Apply the same ideas used in solar thermal power plants (although on a much smaller level), and you're on your way to year-round growing. Solar thermal greenhouses, also called active solar greenhouses, require the same basics as any other solar thermal system: a solar collector, a water storage tank, tubing or piping (buried in the floor), a pump to move the heat-transfer medium (air or water) in the solar collector to storage, and electricity (or another power source) to power the pump.

In one scenario, air that collects in the peak of the greenhouse roof is drawn down through pipes and under the floor. During the day, this air is hot and warms the ground. At night, cool air is drawn down into the pipes. The warm ground heats the cool air, which in turn heats the greenhouse. Alternatively, water is sometimes used as the heat-transfer medium. Water is collected and solar-heated in an external storage tank and then pumped through the pipes to warm the greenhouse.

SMOKELESS SMOKESTACKS: SOLAR THERMAL CHIMNEYS

Just as solar thermal greenhouses are a way to apply solar thermal technologies to an everyday need, solar thermal chimneys, or thermal chimneys, also capitalize on thermal mass materials. Thermal chimneys are passive solar ventilation systems, which means they are nonmechanical.

Examples of mechanical ventilation include whole-house ventilation that uses fans and ducts to exhaust stale air and supply fresh air. Through convective cooling principles, thermal chimneys allow cool air in while pushing hot air from the inside out. Designed based on the fact that hot air rises, they reduce unwanted heat during the day and exchange interior (warm) air for exterior (cool) air.

Thermal chimneys are typically made of a black, hollow thermal mass with an opening at the top for hot air to exhaust. Inlet openings are smaller than exhaust outlets and are placed at low to medium height in a room. When hot air rises, it escapes through the exterior exhaust outlet, either to the outside or into an open stairwell or atrium. As this happens, an updraft pulls cool air in through the inlets.

In the face of global warming, rising fuel costs, and an ever-growing demand for energy, energy needs are expected to increase by nearly the equivalent of 335 million barrels of oil per day, mostly for electricity. Whether big or small, on or off the grid, one of the great things about solar thermal power is that it exists right now, no waiting. By concentrating solar energy with reflective materials and converting it into electricity, modern solar thermal power plants, if adopted today as an indispensable part of energy generation, may be capable of sourcing electricity to more than one hundred million people in the next twenty years. All from one giant renewable resource: the sun.

HOW SPRAY-ON
SOLAR PANELS WORK

I f you've ever used a solar-powered calculator, you've experienced the power of thin-film solar cells. But can spray-on solar panels take that technology one step further?

1 HOUR OF SUNLIGHT = 1 YEAR OF ENERGY

Solar power seems easy enough to use—there's plenty of sunlight. In fact, the sun provides Earth with enough solar energy in one hour (4.3×10^{20} joules) to power all of our energy needs for one year (4.1×10^{20} joules). But the dilemma over the years has been how to harness that solar energy efficiently and put it to use.

Traditional solar panels, the kind you see on rooftops, are crystalline silicon PV arrays—solar panels made up of a collection of solar cells. More recently, thin-film solar technology has become the darling of the solar industry.

These solar cells are made with copper indium gallium selenide, or CIGS technology, and unlike the rigid panels, they're flexible and can be used in places other than rooftops (on windows, sides of buildings, cars, computers, and so on). Affordable, abundant solar technology is coming soon to you.

SPRAY-ON SOLAR PANEL EFFICIENCY

Current commercial PV solar technologies rely on solar cells that are made of silicon that's been coated with a thin layer of silicon nitrate. (The silicon nitrate works as an antireflective material to increase the cell's sunlight-collecting efficiency.) They're costly to manufacture for two reasons: they use hydrogen plasma to collect sunlight, and they are made in a vacuum. Thin-film PV cells use cheaper materials but are more complex to make—and despite the cheaper materials, the production complexity equals a more expensive end product.

Enter the spray-on solar material project. Researchers are experimenting with ways to change how solar cells are manufactured, as well as how to increase solar cell efficiency. For example, a team at Australian National University developed a new method that involves spraying solar panels as they roll down a conveyor belt during production, first with a hydrogen film and then an antireflective film.

Solar cells are made from semiconducting nanoparticles called quantum dots. These quantum dots are mixed with a conducting polymer to make a plastic. Spray-on solar panels composed of this material can be manufactured to be lighter, stronger, cleaner, and generally less expensive than most other solar cells in production today. They are the first solar cells able to collect not only visible light but infrared waves, too.

The Australian team, in collaboration with the German solar company GP Solar, also studied how to increase the efficiency of the cells. Researchers are exploring how the surface of a solar cell (specifically, its roughness) affects its ability to collect solar energy. Right now, the efficiency rate of solar cells on the market is about 15 percent, and the German Fraunhofer Institute for Solar Energy recorded a new efficiency record of 44.7 percent in September 2013. By comparison, the first solar cells manufactured in the 1950s converted less than 4 percent of collected solar energy into usable power. Scientists predict they may be able to increase that rate by five times the current numbers.

USES FOR SPRAY-ON SOLAR PANELS

Solar panel efficiency, fabrication technology, and manufacturing engineering are important not only for the solar industry but also for you, the consumer. New technology and inexpensive materials and production mean more practical, everyday applications.

Currently, applications of traditional commercial PV solar panels and solar-energy systems are too pricey for most of us, aside from affixing rigid solar panels to the rooftops of our homes.

PV technology is used to power spacecraft, bring electricity into remote villages in developing countries, and power remote buildings (or anything that requires electricity).

PICTURE A CALCULATOR

If you've ever used a solar-powered calculator, you've experienced the power of thin-film solar cells. The cell's flexible nature allows it to go places where traditional panels can't, including into private homes and electronic devices, but it's also used in similar energy-producing ways on buildings and in remote locations.

Spray-on solar panels will be sold as a hydrogen film that can be applied as a coating to materials—potentially everything from a small electronic device to a new way to power an electric car's battery. Similar to the solar technology of today, spray-on panels could be incorporated into buildings themselves, not just rooftops. One day you may buy clothing with solar film woven into the fabric.

Perhaps the biggest marketing hurdle facing both spray-on solar panels and the solar industry as a whole is cost-effectiveness. Investing in solar energy research and new solar energy systems is expensive, and soaring expense is a barrier to adopting new technologies.

COOL OFF IN THE SUN: HOW SOLAR AIR CONDITIONERS WORK

olar air conditioners take advantage of the sun at its brightest and use its energy to cool us during the hottest part of the day. What are we waiting for?

American homeowners spend more than $15 billion on home cooling, and roughly 8 percent of all the electricity produced in the United States is consumed by conventional air-conditioning units. All that cooling releases an estimated 195 million tons (177 million metric tons) of carbon dioxide (CO_2), a greenhouse gas known to contribute to climate change.

Greenhouse gas emissions have increased by 25 percent in the 150 years since the Industrial Revolution. In 2010, fossil fuels supplied 81 percent of the energy consumed in the United States.

Demand for air conditioning is expected to grow as temperatures increase. Heat waves cause blackouts, health problems, and in some cases, even death. During the summer of 2003, for example, at least thirty-five thousand people died from a heat wave that baked Europe. Seven of the eight hottest years on record have happened since 2001; the ten hottest years have all been since 1995. Global surface temperatures have increased by 1.4°F (0.8°C) since 1920, and scientists predict temperatures could increase an average of 2 to 11.5°F (1.1 to 6.4°C) by the end of the twenty-first century.

Reducing the use of fossil fuels and replacing them with renewable resources for energy is vital to slowing down the effects of climate change. The only option for reducing the energy consumed by air conditioning has been to simply turn it off—until now. Solar air conditioners take advantage of the sun at its brightest and use its energy to cool us during the hottest part of the day.

COOLING COSTS: THE ENVIRONMENTAL BENEFITS OF SOLAR AIR CONDITIONERS

Conventional air conditioners running at the hottest points of the day contribute to power-grid demands that often lead to outages. Solar air-conditioning units offer environmental benefits including lower grid demand and load shifting during peak usage, reduced electricity costs, fewer power outages, off-the-grid capabilities, and reduced greenhouse gas emissions.

Solar air-conditioning units come in two basic forms: hybrids and chillers.

- A **hybrid system** combines PV technology with direct current (DC). It automatically switches between solar and battery power as needed. When it's set to hybrid mode, the system charges its batteries when the sun is shining. When the sun isn't out, the system runs on battery backup while charging its batteries via alternating current (AC) power.

- **Solar-powered absorption chillers**, also known as evaporative coolers, work by heating and cooling water through evaporation and condensation. Chillers cool the air by blowing it over water-saturated material. Solar energy is used to power the fan and motor. SolCool's hybrid solar air conditioner, for example, runs on solar energy. It can be plugged in or run off batteries. Even when plugged in to a conventional power source, it operates at a maximum of 500 watts per hour, compared to about 900 for a conventional window unit (and 3,500 watts for an hour of central air conditioning). Its chiller option offers air conditioning for hours after a power failure.

COOL CONCERNS: THE CHILLS WE MAY GET ABOUT SOLAR AIR CONDITIONERS

For those looking to replace their conventional air conditioning with a greener option, solar-powered absorption chillers offer reduced energy consumption but will increase a home's water consumption. By design, chillers need to be hooked up to a water line or water storage tank, making them a little less green than the hybrid solar-powered air conditioners that run on solar energy and battery power.

And no matter which style you prefer, a solar-powered air-conditioner unit is going to cost you. It's not as much as installing solar panels on your roof, but while medium-sized, conventional window air-conditioning units typically sell in the hundreds of dollars, a solar-powered system will cost you a few thousand plus installation fees. As previously mentioned, chillers will also need to be hooked up to a water line. However, the cost savings to the planet and to your wallet in the long run could be encouragement enough.

Solar power sounds promising, but as we'll see in the next section, we could tap into some other sources of power—including a steamy one.

PRIUS TECHNOLOGY

Toyota redesigned its Prius with a solar-powered ventilation system that keeps the car cool while it's parked—no more returning to a molten-hot car after it's been parked in the sun all day. The system is mounted on the roof (it comes as part of the solar roof package) and is basically a fan powered by a solar panel. The fan circulates air in and out of the cabin and reduces cabin temps to about that of the outside ambient temperature.

LITERARY POWER: COULD STEAMPUNK INSPIRE THE FUTURE OF ENERGY?

A determined young man, hair neatly slicked to each side of a severe center part, wrestles with the equipment he's installing in the cockpit of a rocket. His long-sleeved cotton shirt is tucked into straight-legged linen trousers, and the required suspenders are performing admirably despite the undignified positions his work necessitates. It's his waistcoat that's giving him fits, migrating toward his chin with each turn of the wrench.

He tosses the tool, hears the metallic clang on the control center's floor, and eases himself into the velvety armchair that faces the mammoth machine's panel of directional dials and gesticulating gauges. He'll return to his work in a moment, but for now, he's content to grasp the sticks that control the rocket ship's hydraulic aft rudder and gaze skyward. Before long, that's where he'll be steering his steam-powered invention.

If this scenario seems like an unlikely marriage of nineteenth-century Victorian culture and modern-day technology, that's because it is. The implausible idea that an

1800s-era man, however inventive, could build a rocket ship capable of navigating the stars is just one of the many old-meets-new plots that come together in the literary subgenre known as steampunk.

Steampunk imagines how people in the past might have adopted technology from the future. Instead of relying on future power sources like electricity, the majority of steampunk-style contraptions are powered directly by steam. And so were many actual Victorian-era inventions, such as steam-powered pumps used to remove water from coal mines. These machines operated like all basic steam-powered inventions of the day. The heat from a fuel, such as coal or wood, caused compact liquid water molecules to expand. As the heat triggered the molecules to move away from each other, they were transformed from a liquid state into a vapor state—steam—and the expansion pushed a piston to power the pump.

Despite all the high-tech advances of the last hundred years, steam is still relevant as an energy source. It's used to generate electricity in fuel-burning and nuclear power plants, and has become an important addition in home appliances, like dishwashers and clothes dryers. More importantly, steam's benefits have inspired researchers to take another look at this elemental energy source and decipher how it could be harnessed for the future.

Let's take a closer look at what some steam enthusiasts are dreaming up.

IIIIₗI THE FUTURE OF STEAM ENERGY

Many people believe steampunk is the brainchild of two well-known authors from the late 1800s, H. G. Wells and Jules Verne. Verne's book *From the Earth to the Moon*, published in 1865, depicted submarines, solar sails, and a rocket-like projectile that transported people to the moon. Although these inventions seemed improbable for many years, several of the contraptions he penned have been incarnated in modern machines. Steam has experienced a similar revival. It's far from being a defunct energy source, a timeworn literary construct, or a remnant of the Industrial Revolution.

According to the U.S. Department of Energy, more than 45 percent of the fuel burned by U.S. manufacturers creates steam, and most of the electricity in the United States is created by steam turbines. A basic steam turbine uses fuel to heat a boiler that converts water to steam. A compressor then condenses the steam into a high-pressure mass that's transferred to a spinning turbine, where it generates electricity to power factories, homes, or even vehicles. Steam-powered vehicles are still in development, but when compared with gas-fueled vehicles, they're expected to be more energy-efficient and have fewer environmentally harmful emissions.

Before steam power can take a quantum leap, however, scientists need to overcome a major efficiency issue. Current steam generators don't capture all the steam that's produced. A large percentage of it (up to two-thirds) is lost. Some steam simply escapes into the atmosphere, while some cools and is recaptured as wastewater. While the wastewater is often recycled right back into the steam generator, it would be more efficient to capture it in the first place, which is exactly what researchers are attempting to do by developing highly efficient steam turbines.

One of the distinct advantages of steam power is that nearly any type of fuel can be burned to turn water into steam. Instead of burning fossil fuels to make steam, scientists contend that combustible organic waste materials, like corncobs or soy oil, could be used. Steam turbines could even be heated using waste wood. The Seattle Steam Company, for example, burns used wood pallets, broken tree limbs, and construction scraps to produce enough steam to heat two hundred downtown Seattle buildings.

Water is a nonhazardous, inexpensive, and plentiful natural resource that can generate up to six times its mass in steam, which means it holds promise as a cleaner energy with widespread industrial and residential applications. Even if none of those applications is a steam-powered rocket that can catapult an impeccably dressed human into space—yet.

COINING A PHRASE

The earliest mention of the word "steampunk" occurred in 1987, when author K. W. Jeter proposed the term in a letter to the sci-fi magazine *Locus*. Today, steampunk refers to books like *The Difference Engine*, which chronicles the invention of the first computer in a nineteenth-century setting, as well as comics like *The League of Extraordinary Gentlemen*, films like *Wild, Wild West*, and a style of dress popular among costumers at pop culture conventions.

WHAT ARE WE GASSING ABOUT? HOW GASIFICATION WORKS

As we saw in the previous section, some of the most promising, attention-getting energy alternatives aren't always cutting-edge, revolutionary ideas. We all know about windmills, waterwheels, and steam power, which have been around for centuries. Today, a variety of improvements, including innovative turbine designs, are transforming these ancient machines into cutting-edge technologies that can help nations satisfy their energy needs.

There's another old process—one you probably don't know much about—that's gaining in popularity and may join wind and hydropower in the pantheon of clean, renewable energy. The process is known as gasification, a set of chemical reactions that uses limited oxygen to convert a carbon-containing feedstock into a synthetic gas, or syngas.

It sounds like combustion, but it's not. Combustion uses an abundance of oxygen to produce heat and light by burning. Gasification uses only a tiny amount of oxygen, which is combined with steam and cooked under intense pressure. This initiates a series of reactions that produces a gaseous mixture composed primarily of carbon monoxide and hydrogen. This syngas can be burned directly or used as a starting point to manufacture fertilizers, pure hydrogen, methane, or liquid transportation fuels.

Believe it or not, gasification has been around for decades. Scottish engineer William Murdoch gets credit for developing the basic process. In the late 1790s, using coal as a feedstock, he produced syngas in sufficient quantity to light his home. Eventually, cities in Europe and America began using syngas—or "town gas" as it was known then—to light city streets and homes. But then natural gas and electricity generated from coal-burning power plants replaced town gas as the preferred source of heat and light.

Today, with a global climate crisis looming on the horizon and power-hungry nations on the hunt for alternative energy sources, gasification is making a comeback. The Gasification Technologies Council expects world gasification capacity to grow by more than 70 percent by 2015. Much of that growth will occur in Asia, driven by rapid development in China and India. But the United States is embracing gasification, as well.

Let's take a closer look at how this process works. We're going to start with coal gasification, the most common form of the process.

‖‖‖ FROM CHARCOAL TO CLEAN COAL: COAL GASIFICATION

The heart of a coal-fired power plant is a boiler, in which coal is burned by combustion to turn water into steam. The following equation shows what burning coal looks like chemically: $C + O_2 = CO_2$. Coal isn't made of pure carbon, but of carbon bound to many other elements. Still, coal's carbon content is high, and it's the carbon that combines with oxygen in combustion to produce carbon dioxide, the major culprit in global warming. Other by-products of coal combustion include sulfur oxides, nitrogen oxides, mercury, and naturally occurring radioactive materials.

The heart of a power plant that incorporates gasification isn't a boiler, but a gasifier, a cylindrical pressure vessel about 40 feet (12 meters) high by 13 feet (4 meters) across. Feedstocks enter the gasifier at the top, while steam and oxygen enter from below. Any kind of carbon-containing material can be a feedstock, but coal gasification, of course, requires coal. A typical gasification plant could use 16,000 tons (14,515 metric tons) of lignite, a brownish type of coal, daily.

A gasifier operates at higher temperatures and pressures than a coal boiler—about 2,600°F (1,427°C) and 1,000 pounds per square inch (6,895 kilopascals), respectively. This causes the coal to undergo different chemical reactions. First, partial oxidation of the coal's carbon releases heat that helps feed the gasification reactions. The first of these is pyrolysis, which occurs as coal's volatile matter degrades into several gases, leaving behind char, a charcoal-like substance. Then, reduction reactions transform the remaining carbon in the char to a gaseous mixture known as syngas.

Carbon monoxide and hydrogen are the two primary components of syngas. During a process known as gas cleanup, the raw syngas runs through a cooling chamber that can be used to separate the various components. Cleaning can remove harmful impurities, including sulfur, mercury, and unconverted carbon. Even carbon dioxide can be pulled out of the gas and either stored underground or used in ammonia or methanol production.

That leaves pure hydrogen and carbon monoxide, which can be combusted cleanly in gas turbines to produce electricity. Or, some power plants convert the syngas to natural gas by passing the cleaned gas over a nickel catalyst, causing carbon monoxide and carbon dioxide to react with free hydrogen to form methane. This "substitute natural gas" behaves like regular natural gas and can be used to generate electricity or heat homes and businesses.

But if coal is unavailable, gasification is still possible. All you need is some wood.

SYNGAS SECONDS

Although the electric power industry has recently become interested in gasification, the chemical, refining, and fertilizer industries have been using the process for decades. That's because the major components of syngas—hydrogen and carbon monoxide—are the basic building blocks of several other products. Some of the most important products derived from syngas include methanol, nitrogen-based fertilizers, and hydrogen for oil refining and transportation fuels. Even slag, a glasslike by-product of the gasification process, can be used in roofing materials or as a road-bed material.

THE ORGANIC OPTION: WOOD GASIFICATION

Coal gasification is sometimes called "clean coal" because it can be used to generate electricity without belching toxins and carbon dioxide into the atmosphere. But it's still based on a nonrenewable fossil fuel. And it still requires mining operations that scar the earth and leave behind toxic wastes of their own. Wood gasification—or biomass gasification, to be more technically correct—may provide a viable alternative. Biomass is considered a renewable energy source because it's made from organic materials, such as trees, crops, and even garbage.

Biomass gasification works just like coal gasification. A feedstock enters a gasifier, which cooks the carbon-containing material in a low-oxygen environment to produce syngas. Feedstocks generally fall into one of four categories:

- **Agricultural residues are left after farmers harvest a commodity crop.** They include wheat, alfalfa, bean or barley straw, and corn stover. Wheat straw and corn remnants make up the majority of this biomass.
- **Energy crops are grown solely for use as feedstocks.** They include hybrid poplar and willow trees, as well as switchgrass, a native, fast-growing prairie grass.
- **Forestry residues include any biomass left behind after timber harvesting.** Deadwood works well, too, as do scraps from debarking and limb-removal operations.
- **Urban wood waste refers to construction waste and demolition debris that would otherwise end up in a landfill.** Pallets—flat transport structures—also fall into this category.

The choice of feedstock determines the gasifier design. Three designs are common in biomass gasification: updraft, downdraft, and crossdraft. In an updraft gasifier, wood enters the gasification chamber from above, falls onto a grate, and forms a fuel pile. Air enters from below the grate and flows up through the fuel pile. The syngas, also known as producer gas in biomass circles, exits the top of the chamber. In downdraft or crossdraft gasifiers, the air and syngas may enter and exit at different locations.

The choice of fuel and gasifier design affects the relative proportions of compounds in the syngas. For example, wheat straw placed in a downdraft gasifier produces the following:

- 17 to 19 percent hydrogen gas
- 14 to 17 percent carbon monoxide
- 11 to 14 percent carbon dioxide
- virtually no methane

But charcoal placed in a downdraft gasifier produces the following:

- 28 to 31 percent carbon monoxide
- 5 to 10 percent hydrogen gas
- 1 to 2 percent carbon dioxide
- 1 to 2 percent methane

Now you're ready to make your own wood gasifier. Keep reading to see how.

BETTER GAS(IFICATION) MILEAGE

Believe it or not, one of the main uses of wood gasification has been to power internal combustion engines. Before 1940, gasification-powered cars were occasionally seen, especially in Europe. Then, during World War II, petroleum shortages forced people to think about alternatives. The transportation industries of Western Europe relied on wood gasification to power vehicles and ensure that food and other important materials made it to consumers.

After the war, as gas and oil became widely available, gasification was largely forgotten. A future petroleum shortage, however, may revitalize our interest in this old technology. The car driver of the future may ask to "fill 'er up" with a few sticks of wood instead of a few gallons of gas.

▐▐▐▐ DIY: HOMEMADE GASIFICATION

One attractive quality of gasification is its scalability. The Polk Power Station just southeast of Tampa is a gasification plant covering 4,300 acres (1,740 hectares). It converts 100 tons (90.7 metric tons) of coal an hour into 250 million watts of power for about 60,000 homes and businesses.

But you don't have to be a giant public utility to experiment with gasification. You can build a simple, small gasifier with materials you find around the house. YouTube features several videos of these homemade units. One video, for example, shows a paint can playing the role of the pressure vessel in which gasification reactions occur. As the syngas is produced inside the sealed can, it moves through some simple plumbing fittings to a burner can, where the gas can be ignited.

Another interesting video shows a small team assembling and operating a wood gasifier based on plans prepared by the U.S. Federal Emergency Management Agency (FEMA) and the Oak Ridge National Laboratory. FEMA developed these plans in 1989 specifically for small-scale gasification in the event of a petroleum emergency. The agency's report includes detailed, illustrated instructions for the fabrication, installation, and operation of a downdraft biomass gasifier.

The unit requires a galvanized metal trash can, a small metal drum, common plumbing fittings, and a stainless-steel mixing bowl and can be mounted on a vehicle to provide syngas for internal combustion. With the gasifier in place, the vehicle can run reliably using wood chips or other biomass as the fuel.

If you're interested in gasification, but aren't the do-it-yourself type, then you might want to consider buying a gasification unit from a manufacturer. For example, New Horizon Corporation distributes gasification systems that can be installed in a home environment. These biomass gasification boilers can heat houses, garages, and other buildings and can use a variety of fuels, including seasoned wood, corncobs, sawdust, wood chips, and any kind of pellet.

Either way, gasification will likely emerge as one of the most important energy alternatives in the coming decades. It's the cleanest way to use coal but also works efficiently with renewable energy sources, such as biomass. And, because one of the primary products of gasification is hydrogen, the process offers a stepping-stone to producing large quantities of hydrogen for fuel cells and cleaner fuels.

HOLY HYDROGEN: COULD HYDROGEN BE THE FUEL OF THE FUTURE?

As we've seen, when it comes to choosing the energy source that will replace fossil fuels, there is no shortage of options. But hydrogen stands apart as a promising alternative energy source. Although the idea of hydrogen as a widely used fuel source to power cars and generate electricity is a relatively new concept in response to seeking an alternative to oil, hydrogen fuel cells actually predate the international combustion engine, which was invented in the middle of the nineteenth century, by about twenty years.

Given that the most basic form of this technology has been around for nearly 150 years, why has its time suddenly come?

WHY HYDROGEN?

Hydrogen is the most abundant element in the universe, so there's no chance of human consumers depleting the supply. There are certainly enough oil resources to meet global demand now, but many energy experts predict that the world's supply of oil will be depleted within sixty years. Hydrogen is so easy to produce that the process could be completed at home with the right equipment.

Exhausting the world's supply of oil or even approaching the inevitable shortfalls that come with a growing population—the planet now hosts seven billion people as of this writing—and economic growth will not only create a major energy crunch necessitating the rapid introduction of alternative energy sources, but burning that much fuel also means an enormous burden on the atmosphere. And that doesn't even account for the potential environmental consequences of extracting crude oil from the earth.

By contrast, hydrogen is clean-burning. The only by-product of hydrogen power is water and heat, both of which can be recycled. This essentially means turning an energy-consuming process into an energy-producing one.

Hydrogen, however, is not a ready source of energy like oil and natural gas. Rather, it is a means of storing energy, since pure hydrogen isn't available on Earth in quantities necessary to fuel an entire energy economy. To get hydrogen in the form of a usable fuel requires energy. Hydrogen can be produced either by separating it from oxygen molecules in water through the process of electrolysis, or by splitting it off hydrocarbon chains in fossil fuels, a process that itself creates greenhouse gas emissions.

FROM H-BOMBS TO HOUSES: HOW DOES HYDROGEN GENERATE USABLE POWER?

Hydrogen fuel cells don't work quite like petroleum-based combustion engines, which rely on heat and power to create energy.

A fuel cell is made up of a stack, "a sandwich of anodes, cathodes, and other high-tech materials," as HowStuffWorks .com's Ed Grabianowski explains. Liquid hydrogen fuel enters around the anodes, where electrons attached to the hydrogen are separated from the atoms. An electrolyte within the fuel cell allows hydrogen protons to pass through, but not the electrons. When the hydrogen atoms reach the other side of the fuel cell, the cathode, it binds with oxygen, creating heat and water vapor.

HYDROGEN HASSLES: WHAT ARE THE DRAWBACKS OF HYDROGEN POWER?

Unless you happen to live in California, which has taken some initiative in building the infrastructure to support hydrogen fueling, chances are you've never seen the option of filling your car up with hydrogen at the gas station.

If you've been to a car dealership recently, chances are you've seen vehicles with ordinary gasoline engines, diesels, and probably even a few hybrids. But hydrogen-powered cars? Not likely.

And this is one of the biggest hurdles to implementing a new energy technology: energy producers and distributors need the infrastructure to supply demand for their fuel. But the demand cannot really exist without the infrastructure to support it. It's like the "chicken and the egg" problem, but the difference is that the solution is probably worth billions.

Another significant drawback of hydrogen is that, although it's abundant, hydrogen fuel can be difficult and costly to store. At normal room temperatures, hydrogen exists as a gas. To get hydrogen into a liquid state that can be stored, transferred, and eventually used as fuel requires a temperature of −423°F (−253°C). Keeping hydrogen fuel that cold requires specialized containers.

Finally, hydrogen-powered cars are currently too expensive for the average consumer to purchase. Toyota announced that by 2015 it will produce a hydrogen-powered vehicle, but did not say how much it would cost. At one point, production costs of each vehicle ran as high as $1 million.

In other words, while hydrogen is promising and has considerable attention and investment from energy companies and auto manufacturers alike, its time hasn't yet come. So yes, hydrogen may be the fuel source of the future, but tomorrow, you'll still probably need to fill your car with regular old gasoline.

FARTING MICROBES AND MORE: FIVE EMERGING TECHNOLOGIES IN THE ENERGY INDUSTRY

At this point, we've heard about some of the biggest potential powers in the world of energy: solar, steam, clean coal/gasification, and hydrogen. Old-school energy technologies, like coal, natural gas, and oil, just aren't going to cut it anymore, cost-wise or environment-wise. And the new-school technologies aren't quite ready to power everything from our smartphones to our cars. So what do we do in the meantime?

That's where these next ideas come in. This list has it all: farting microbes, pollution-reducing oxides that sound like terrible hairdos, and even jails with their own self-sustaining grids! (It is a bit of a relief to know that when the power gets knocked out everywhere else, it might not in the jails.)

So the old stuff gets a makeover, and the new stuff gets ready for prime time. And all of it is either here already or will be on its way shortly.

1. STREAMLINE THE TURBINE

Some people look at a field of twirling, white wind turbines and see a cleaner, brighter future. Some people look at those turbines and see whirling blades of death for birds.

Enter the Saphonian, which is not the bad guy in a new *Star Trek* movie, though it should be. It is a bladeless wind turbine. It looks a bit like a satellite dish on a stalk rather than like a windmill designed by IKEA. Besides not killing birds, the Saphonian is easier on the ears of humans and animals alike and converts wind power to energy more efficiently. It's a win-win. Or a wind-wind, if you like puns.

IIIII 2. HARNESSING THE MOTION OF THE OCEAN

You know what they say: it's not the size of the wave, it's the motion of the ocean. Turns out that's pretty much true. And now Maine is putting it to the test with the first tidal generator to go online and actually power things.

Almost every state with a coastline has dabbled in capturing the energy of water waves, but the Bay of Fundy is making it a reality with the Maine Tidal Energy Project.

The first turbine delivered electricity to the U.S. power grid in September 2012, and the plan is for it to generate 150 kilowatts for the grid as water runs through the turbine at about 7 miles per hour (11.3 kilometers per hour). Once all the bugs are worked out of the first turbine, the company hopes to set up twenty more and deliver about 3 megawatts of power when they're all hooked together. That's enough to power 1,200 homes. Not much, but it's a start.

3. GET YOUR OWN GRID

Some people, often in hemp pants and vegan shoes, want to get off the grid and live on the land in harmony with nature. Or maybe they want to stick it to the man. Or maybe they're in a maximum-security jail.

Santa Rita Jail in Alameda, California, near San Francisco, uses an array of fuel cells, solar panels, wind turbines, and diesel generators to power its very own micro grid. All the jail's power is generated on-site, which means it doesn't have to connect to a central power plant or be on the grid at all. When a storm knocks out power in Alameda, the electricity at the jail is still on. That's a relief, eh?

In addition to safety, the micro grid delivers cleaner power for less money. The jail actually saves $100,000 a year and sells some of its excess power back to the grid. And because the solar panels and wind turbines are hooked up to batteries and diesel generators, the fickle nature of sun and wind don't cause dips in power. They've got backup.

4. BUG FART POWER!

Is there anything microbes can't do? It seems like every time we turn around, they're causing epidemics or doing good in our guts or solving quantum equations. Sure, maybe they don't do higher mathematics, but they do fart methane. And we like that. Really.

These gaseous little guys, some of whom live at Stanford and Penn State with their scientist friends, are called methanogens. It only sounds like they should have a cameo on *Breaking Bad*. They do not make drugs. In the wild, if you can imagine free-range methanogens, they would eat carbon dioxide from the air and electrons from hydrogen gas and excrete pure methane.

The scientists have figured out that if the methanogens are fed a stream of electrons from emissions-free power sources, like solar cells and wind farms, to go with their main dish of CO_2, the methane emissions that result can be turned into fuel for airplanes, ships, and cars.

5. DEPOLLUTE DIESEL

For a long time, diesel had a terrible reputation. And it deserved it. It was stinky and filthy, and sent out noxious black clouds from the tail pipes of nasty little cars. Sure, it got more miles to the gallon than gasoline, but was it worth it when you could see and smell the pollution?

Over the years, engineers and chemists have worked to clean up diesel's reputation, and the latest in clean diesels are a vast improvement over the old stuff. But scientists at the

University of Texas at Dallas think they've found a way to reduce the pollution and the cost of cleaning up diesel.

Since diesel fuel currently uses platinum—that's right, the stuff that hip-hop stars' dreams are made of—to reduce pollution, using just about anything else would make it cheaper. The new material is a business-up-front, party-in-the-back man-made material called mullite. Not only is it cheaper, as you'd expect from an oxide called mullite, but it also reduces pollution from diesel fuel by 45 percent.

3

HOME IS WHERE THE ROBOT IS:
THE INTELLIGENT FUTURE OF OUR HOUSES, HIGHWAYS, AND HOVERCRAFTS

S o far, we've looked at how we'll learn and have fun in the future, how we'll heal our minds and bodies, and how we'll power our increasingly high-tech world. But what about how we'll house ourselves? Or even more important, how we'll get around? Let's explore our future environment, starting with the aesthetics of our future abodes.

As much as we'd all love to hang our hats in *Jetsons*-style bubble homes in the future, it's likely that future domiciles will be more practical than fantastical.

The future of architecture seems

to have two main prongs: sustainable design and the sleek, high-tech look. At first glance, these two directions may seem to be mutually exclusive. For some people, green living conjures up visions of existing close to the earth—houses built out of straw by their owners, with rain barrels to water organic gardens and turbines to harness wind power. Crunchy, hippie, granola-eating stuff, and very low-tech.

On the other hand, a high-tech home brings to mind geeks who are big into electronics—more like *The Jetsons* in a lot of ways. The truth is that the future of architecture incorporates both types of elements—the minimalist, modern, sleek aesthetic and the environmentally friendly, money-saving practicality. And while homes will probably always have the same basic features (a roof, windows, a kitchen, a bathroom, a living room), the future of architecture has the power to change the way that we live—for the better.

BUILDING UP, NOT OUT

According to Denis Hayes, one of the founders of Earth Day, a whopping 82 percent of Americans, and over half of all human beings on the planet, dwell in cities, which weren't designed with sustainability in mind.

In reality, Earth is running out of resources and room. At least in the places where many of us want to live, meaning the cities. In large cities, this means that people who can afford it often pay big bucks for small spaces, while people who live in the suburbs and work in the cities spend time, money, and gasoline commuting to their jobs.

While there are lots of potential solutions to these problems, types of architecture can make a big difference. One option is the super-tall building—not just a skyscraper, or high-rise as we think of them, but nearly 1,000 feet (300 meters) high or even more. It's truly vertical living. We've already seen a trend toward more mixed-use communities—meaning living, playing, shopping, and working all in one area—with promising results. The super-tall building trend takes this to the next level, since the sky is literally the limit.

Imagine how living in a building like this could change your life. Your coworkers could be your neighbors. (Hopefully you like them in both settings.) You'd feel more invested in your environment since it's all contained in one place. The idea is to not only have a smaller physical footprint on Earth but also build a lively, dynamic community of people. No more urban sprawl. There are even ways to lessen these über-tall buildings' impact on the environment, such as using electronic glass panels that darken when exposed to higher temperatures to help cool the building and absorb sunlight to generate energy.

Buildings are already being designed with this ideal in

mind. The iceberg-shaped London edifice known as The Shard has seventy-two floors that include office space, residences, shopping, and more. Its builders also boast that it is more energy-efficient and has indoor gardens.

Don't want to live in a big, shiny tower? Going more compact is also the way of the future for single-family homes, for both economic and environmental reasons.

SMALLER IS BETTER

Owning a single-family home will still be a goal for many of us in the future. But according to the National Association of Home Builders, those houses are going to be smaller, and features like formal living rooms will disappear. The trend is toward more multipurpose, open living spaces.

In 2012, the average new home size in the United States was about 2,306 square feet (214 square meters). A survey published in September 2011 revealed that about 32 percent of the respondents preferred a home that's 1,400 to 2,000 square feet (130 to 186 square meters). The heyday of McMansions appears to be over. Can smaller be better? In this age of flat-screen TVs and Kindles, how much room do we really need for all of our stuff?

Contests like the U.S. Department of Energy's Solar Decathlon provide a possible glimpse at these homes of the future. College teams design and build homes that are energy-efficient, solar-powered, and affordable. The 2011 winning team, from the University of Maryland, was inspired by the ecosystem of the Chesapeake Bay. Their house, known as WaterShed, is modular and has the lines and angles we've come to expect from a "future house." Its split, butterfly-winged roof is designed to collect rainwater in a central core, and it features a garden of native plants, a composting system, and an edible wall. What about the technology? WaterShed also has solar thermal arrays on its roof and an automated system to regulate the house's thermostat and lighting. It also blends very nicely into its environment.

Sweet, you may be thinking, but I want my house to look all futuristic! No worries. There's no one set "architecture of the future." For example, a company called Alchemy Architects is already selling prefabricated homes called weeHouses. They comprise modules that look like shipping containers, and the houses can be as small as one container (435 square feet or 40 square meters) or as large as four (1,765 square feet or 164 square meters).

You work with the company to design the right one for you, buy your home site and get it ready with a foundation and utilities, and Alchemy ships out your house. Appliances and green features like bamboo flooring and alternative energy sources are all available, and the company claims that because they're prefab, its houses are less expensive than conventional homes.

While those highly creative, futuristic architectural designs are fun to check out, the real future of architecture—especially for us nonmillionaires—will probably be more subtle and practical. Smaller, more affordable, greener, and yes, with even more advances in technology. In the next section, we'll predict what our individual residences will look like in the next fifty to a hundred years.

REAL-ESTATE REALITY CHECK: WHAT WILL HOMES LOOK LIKE IN FIFTY TO ONE HUNDRED YEARS?

Fifty years ago, people thought "homes of the future" would look something like wild, bubble-shaped pod houses where humans floated from room to room, trailed by robotic help. In other words, a far cry from how modern homes really turned out. There are no pod houses, no floating from room to room, and no robot servants (unless you count the Roomba). Most of us live in homes that are very traditional, and styles of architecture from the past still have cult followings.

Even homes that are more cutting-edge have many of the same features that homes did fifty to one hundred years ago. We still have kitchens, living rooms, bedrooms, and bathrooms. We're still sitting on sofas, eating at tables, and sleeping on mattresses resting on wooden or metal bed frames. So will future homes finally meet our wildest futuristic expectations? We make our own predictions in this section.

When thinking about what houses might look like in fifty to one hundred years, it's hard to imagine that they'd be totally unrecognizable to us. It's fun to speculate and admire some of the more unusual, futuristic home styles, but most of them aren't going to become the norm. In addition, some architects today argue that newer houses aren't built to last the way that older homes were. That means that there's a good chance houses in the future will be a mixture of newer architecture (possibly built of longer-lasting materials) and houses standing today that have been upgraded for modern sensibilities.

Let's take a look at what homes might look like in the future, starting with the high-tech and energy-efficient features that have the most staying power.

THE WORDS OF THE DAY ARE TECH-Y *AND* EFFICIENT

It seems a given that most of our electronics and appliances will only get better and more efficient in the future. Although we have plenty of them in our houses right now, getting all of them to work together seamlessly still isn't common for most of us. You can use your smartphone to do everything from watch TV to check on your home security system, but can you also turn on your oven? The home of the future will have robot servants—they just won't be big, shiny metal androids rolling around and cracking jokes. They'll be hidden within your home, unobtrusive, intuitive, and easy to use, and they'll incorporate more sensitive and touch-screen technology.

Microsoft's Future Home, for example, is full of this kind of tech. It's a concept home that the company updates periodically to reflect what it thinks the future will hold, but so far, many of its concepts seem right on track. It doesn't seem like much of a stretch to go from a mat by your front door that charges all of your electronic devices to one that can also retrieve information from those devices and display it on a big, touch-sensitive screen.

Challenges for incorporating this kind of technology today include the expense of retrofitting an existing home, but future houses will incorporate it just like they do plumbing and electricity. There are other challenges beyond cost, though. Microsoft also conducted a study to find out how people felt about home automation, and one of the biggest concerns was security. If everything's automated, you have to make sure nobody can hack in and unlock your doors!

Right now environmental friendliness is a huge trend in home building, and you may already have things like energy-efficient windows. These kinds of features will probably become more common, but that doesn't necessarily mean that in fifty years, your house will be covered in solar panels or surrounded by wind turbines. Futurist Ian Pearson argues that solar panels make much more sense in the African

Sahara, for example, than on the roof of a suburban home that sees far fewer sunny days.

Given the fact that all renewable sources of energy have their problems, it's hard to predict what will power your house. But it'll be designed to use that power as efficiently as possible in heating, ventilation, and air conditioning (HVAC) systems. Things like gray water recovery systems will become inexpensive and easy to use, so using your shower water to flush the toilet won't be a big deal at all.

A stealth-tech, eco-friendly home is probably on the horizon, but will the walls themselves still be wood and drywall?

CONCRETE CASTLES, FUNGUS DWELLINGS, AND TREE HOUSES? POSSIBLY!

Buildings really used to be built to last; there are churches in Europe that have been around for thousands of years. But that's often because the only materials available were ones that were naturally durable, like stone. Materials like wood, drywall, fabric, and plastic are less expensive, lightweight, and easier to work with. But one of the downsides is that very few parts of a modern home are built to last more than fifty years. The old materials end up in a landfill, and then energy is spent to make and install new stuff.

Here's where the green movement comes in. Not only do we want houses that are built to last, but we also want ones that are constructed of materials that aren't as taxing on the environment. And we want them to be relatively inexpensive. Wood is renewable, but issues like global warming and mass deforestation have some people looking for alternatives.

MEET THE MEAT HOUSE

A Terreform ONE research group headed by Mitchell Joachim is exploring the concept of a habitat made from pig cells grown in a lab. It makes sense on one hand—the organic structure of bones, muscles, and skin is a pretty amazing one. Joachim even suggests that sphincter muscles could open and close to serve as walls and windows. But it also sounds gross. Would the structure itself be alive, then? Would it smell like bacon? It's a good way to get us thinking about organic building materials, though.

There are some pretty wild concepts in the works, but a new one that seems likely to become a reality is concrete. Concrete itself has been around for a long time, but in the past decade or so, there have been some huge advances in what's called high-performance concrete.

It's much stronger than the traditional stuff, and it's an incredible insulator. It can come in lots of different colors, and it can be poured in just about any shape that you want. Companies that manufacture this material continue to reduce the environmental impact of the process. High-performance concrete lasts a long time and doesn't need to be painted or maintained. It's expensive now, but again, the cost will come down over time.

A concrete home isn't as fun to contemplate, though, as one made out of fungus. Members of the nonprofit design group called Terreform ONE come up with creative green building ideas. Mycoform, for example, is a building block made from a fungus that has been grown in molds, feeding off agricultural by-products like hay. A much more palatable concept, called the Fab Tree Hab, includes growing houses by grafting trees and other plants onto a frame. Indoor, organic structures called living walls are already in use and are sort of a step above your average houseplant. Not only are they pretty, but they also filter the air.

Don't want to live in a fungus or tree house? Expect to see more recycled metals and other materials in use both outside and within the walls of homes in the future.

TOMORROW'S APARTMENTS

We won't all want to live in a single-family home for a bunch of reasons, and the maintenance required to keep up the house and yard is just one of them. But expect the generic rows of apartment buildings to become less common. People want to live where they work and have a sense of community. Common and outdoor spaces mean you don't need as much inside.

Flashy home designs of the future are fascinating, but they probably won't become commonplace. In fifty to one hundred years, we'll probably live in more high-tech, more environmentally friendly, longer-lasting, and smaller homes that look much the same as your home does right now. In fact, some of our houses will be so high-tech, they could be considered truly "aware," as we'll see in the next section.

IT KNOWS WHERE YOU LIVE:
HOW THE AWARE HOME WORKS

Imagine a house that knows where you are at any given moment. One that can pass messages between family members, streamline the tedious process of uploading and preparing home movies and photo collections, and tell you how much energy you're consuming—maybe actually saving you a bit on the power bill. What if it could tell you what the weather's like outside, call an ambulance if you get hurt, remind you if you forget to take your medicine, or help you find that darn remote?

Sound too good to be true? These are just some of the features homes of the future might have. Different technologies that could be used include RFID tags, wearable computing devices, home networking systems, gesture technology, and LCD touch panels. Homes could be flawlessly integrated, embedded with technology, and capable of operating independently of direct human orders, while providing exactly what we want, an idea known as ubiquitous computing.

Nowadays, many people are waiting until later in life

to start families. Frequently, they have young children right around the time mom and dad start to run out of steam and look for other living options. This kind of monkey in the middle is often referred to as being a part of the sandwich generation, meaning they're responsible for the care of young and old family members at the same time. And they're usually working full time, too. Add to this the fact that their children or parents, or both, might have one of any number of chronic conditions—from diabetes to dementia and autism to Alzheimer's—and dealing with it all can get a little tough.

SOMETHING FOR SENIORS

The "sandwich generation" is just one example of a family type who could make good use of the wonder-house mentioned previously. Elderly people who desire to stay living in their own homes—called aging in place—could benefit from many of those same features.

All of these possible functions of a house—a house aware of its residents, their activities, and their needs—have been the basis of studies at the Georgia Institute of Technology in Atlanta. A residential laboratory allows its researchers to examine how people use technology in the home and how technology—both current and future—can be shaped to cater better to the user and enhance the whole experience.

Where exactly did the idea for the Aware Home originate, and where do researchers hope the project will lead?

IIII| RAISING AWARENESS: BUILDING THE AWARE HOME PROGRAM

Let's start by taking a little closer look at how the Aware Home got started and what the goals of the project are.

The Aware Home Research Initiative commenced in 1988, and the Georgia Research Alliance built the Aware Home through a grant. The project is broader in scope than just that abode, though. Researchers from Georgia Tech not only study and develop their own areas of research, but they also collaborate with other academic and corporate bodies, sharing the knowledge they learn.

Since its beginning, the research initiative has moved forward on a variety of fronts, combining aspects of health care, education, entertainment, security, and, of course, technology. The Aware Home project has drawn together researchers specializing in a wide range of fields such as computer science, psychology, health systems, engineering, architecture, assistive technology, and industrial design.

In the future, the folks involved in the Aware Home project—along with their many collaborators, sponsors, and partners—plan to continue exploring issues related to domestic life and the development of ubiquitous computing in the home. Goals of this research initiative include efforts to find out what technologies people have the ability (and the desire) to use in their everyday lives. They also hope to increase collaborative efforts and improve everyone's understanding of the optimum ways an Aware Home could serve people.

Some concerns have been raised in regards to an Aware Home's potential level of invasiveness into people's lives. Because of this, research is also being conducted regarding privacy and the ethical use of monitoring equipment. Another area of concern that's being looked into is the possibility of security risks. A hacker in your network could wreak havoc on your homelife.

Despite this, enhancing the Aware Home's ability to detect changes in health and instances of slow development could prove increasingly beneficial in our day-to-day lives. Those technologies, along with the development of more advanced robotic companions (who wouldn't want one of these?) have many potential areas for expansion.

Now let's examine how the Aware Home can help when someone opts for aging in place.

⫿⫿⫿ AGING IN PLACE

There comes a time in many people's lives when they need to decide whether they'll enter a nursing home, move in with family and friends, or tough it out at the old homestead. Increasingly, people are choosing to spend their golden years the same way they spent all their other years—living independently in a home of their own. For cases such as these, the Aware Home Research Initiative is helping develop a lot of useful solutions.

Probably the most obvious design elements of a senior's Aware Home would concern the ways it's adaptable for people dealing with chronic conditions. The halls and doorways are wide enough to pop a few wheelies in a wheelchair, and the bathroom comes with handy rails. In case an elevator becomes a priority, there's a section of the house that could easily be converted into an elevator shaft. Don't get your hopes too high: elevator installation is very expensive, but having the potential space ready can cut the cost significantly.

Apart from these physical attributes, an Aware Home could offer several other technologies to assist someone aging in place:

● **Digital family portraits** and **discrete motion sensors** that would allow family members instant access to updated information on their loved one's condition and recent activities.

- A program to **help a person keep track of medicines**, with reminders when it's time to take a prescription, advice on potential drug interactions, and other tips to keep health at a premium level.

- Another possible application would **take photos of the person while he or she is cooking**. That way, if the person's memory is going downhill, he or she can keep better track of the steps in the recipe.

These are just a few of the many ways an Aware Home could be used to help someone live independently and boost their failing memory.

An Aware Home could also be programmed to take environmental readings, such as measuring the temperature in the house and questioning the resident if it seems too hot or too cold. It could track how often a person eats and how mobile the person is. If the worst should happen and an elderly person hurts him- or herself or becomes ill, the house could take the necessary steps to get help.

Now that we've seen some of the ways an Aware Home could enhance the life of an elderly person, let's take a look at how it could be of service to whole families.

IIIIᵢ BLTs, TUNA MELTS, MEATBALL SUBS: THE SANDWICH FAMILIES

In the case of a sandwich household, many of the technologies mentioned previously could prove very useful for them, too. But there have been other potential ideas kicked around that could assist families with children—especially if those children have developmental delays, learning disabilities, or emotional disorders.

- Starting when children are born, their development could be monitored and tracked to alert parents to any potentially concerning behavioral or developmental trends. Children have many milestones in their early years, and knowing which ones to check off the list can be critical if there are issues. Plus, you'll never get a better baby monitor than this house.

- In the same manner that the Aware Home could help with organizing a senior's health care, it could also assist with the therapy of children with disorders such as autism. Keeping track of the progress children make with their parents, as well as any other caregivers and therapists, could be important to successfully monitoring their progress and introducing improvements.

- The Aware Home could also provide a safer environment for a child with asthma by monitoring his or her breathing through something as simple as an MP3 player.

But helping with the kids and knowing (through force-sensitive Smart Floor tiles) who's raiding the fridge for a midnight snack wouldn't be the extent of an Aware Home's usefulness. It could also provide a digital message center, a neat and organized way for everyone to stay current with each other's plans. Or it could be connected to a home security system, plus carbon monoxide and smoke detectors—and emergency services in case one of them goes off.

Raising kids, working full time, and possibly juggling the care of an aging parent doesn't leave a lot of free time. Long searches for crucial items like car keys, wallets, and cell phones can throw a day totally off track. An Aware Home could come through and save the day, guiding you to the lost items in just a few seconds. It could also help save some cash by taking care of gray water reclamation, swinging the blinds open when it's sunny, and flipping off the lights when no one's in the room.

And let's not forget the enhancements an Aware Home could bring to home entertainment. For example, sharing video collections and photo albums on social networking and personal websites is becoming increasingly common—but how much time does uploading, tagging, and captioning everything waste? One of the ideas behind the digital media research is to give people back the time they now spend processing photographic and video memories so they have more time to enjoy having memorable moments in the first place.

NOT JUST HOUSING HYPE: FIVE AWESOME FUTURE HOME TECHNOLOGIES YOU'LL LOVE

In the 1950s and 1960s people used to talk about the home of the future, with automatic sliding panels and robot servants. Alas, our homes still aren't the technological wonders we'd dreamed about. But we certainly have made some advances. Think of amenities like central vacuum, in-home stereo, and programmable thermostats. Not sexy enough? Then read about five truly exciting home technologies that are either available now or in development, meaning they all have a real chance of becoming commonplace.

1. LET THERE BE LIGHT OR NOT

Imagine coming home after work and pushing one button that adjusts the lighting for your entire house. The lights go on in the kitchen and living room, maybe your bedroom and select hallways, and the drapes lower throughout the house. Then the exterior lights begin to blaze. Actually, let's go one step further. You have a control panel in your home, much like your programmable thermostat. You program in your daytime and nighttime lighting preferences, plus those for the weekend, and voilà—no need to mess with your lights again.

You may be thinking, "Of course I'd have to mess with them again. What if I wanted different lighting because I was reading or snuggling with a date?" No worries. The system would be so smart and so all-encompassing that you could preprogram settings for romance, reading, dinner parties, and more.

Various advanced lighting systems are currently available, although they're not widely used because they're an added expense. In the future, though, the hope is that they'll be standard features in homes. And why not? Much like the programmable thermostat mentioned earlier, controlled lighting saves energy and money. It's also a great safety and security feature.

2. POWERING DOWN: TRACKING YOUR HOME'S CARBON FOOTPRINT

Did you ever wonder why we've had smart technology in our cars, but not our homes? Think about it. If you're low on gas, your car tells you. If your tire pressure is low, your car tells you. If you're short on oil, need to check your engine, or are out of windshield wiper fluid, your car tells you. All of this information helps keep your car healthy and you safe.

It also saves you money by taking preventive measures and avoiding costly repairs. But what does your home and its major appliances tell you? Nothing. Powerhouse Dynamics, for one, wants to change this.

The Maine-based company has unveiled the Total Home Energy Management program, which monitors a home's energy use, energy cost, and carbon footprint every single minute. Thanks to its detailed tracking, the system can tell you when your energy use spikes and why, allowing you to moderate consumption.

It also monitors your appliances, letting you know when maintenance is needed so you can avoid pricey repairs, and even clueing you in when an appliance is so old and energy-inefficient that it's cheaper to chuck it and buy a new one. And new features are constantly being added as the company identifies and addresses the needs of its customers. For example, when the defrost cycle in a customer's refrigerator wasn't working properly, the system was modified to provide an alert.

3. BIONIC HELP: ROBOTS IN THE HOME

For years—decades—we've been regaled with tales of housekeeping robots. Remember Rosie from *The Jetsons*? Yet here we are, in the twenty-first century, with only a handful of robotic household helpers. All is not lost, however. Inventors are still hard at work trying to perfect the perfect 'bot.

One prototype unveiled by research scientists in Germany is a one-armed, three-fingered wonder that can pick up items while on cleanup duty, serve drinks to its owners and their guests, and even operate some machines. Numerous sensors prevent it from inadvertently (and painfully) clamping its hand around your arm. While users can direct the robot via a touch screen in its serving tray, it also responds to spoken commands, plus can understand and respond to gestures.

This is just one version of a household robot, of course. There are others in development, and it's anybody's guess which one(s) will actually be developed, marketed, and sold. Or if they'll be priced so that the average person can afford one. But we haven't given up on the concept.

4. POTTY TIME: SMART TOILETS

If you've been to Asia, namely Japan, you've likely been fascinated by their toilets. To those of us in North America, they're quite futuristic. These super bowls are loaded with buttons and gadgets whose function and operation are difficult to figure out—at least to the uninitiated. For starters, the toilets also function as bidets. Long common in Europe and other parts of the world, bidets spray water at you for post-potty cleansing. In these newer toilets, a dryer also kicks in, wafting warm air up toward you, meaning there's no need for toilet paper.

But there's so much more to these wonders, like heated seats and lids that raise—and lower—automatically. Not surprisingly, the latter function is especially appealing to women; they've even been dubbed "marriage savers." Then there are the built-in deodorizers, which remove every trace of our prior activities. And, of course, they self-flush, a toilet feature already found in public restrooms in America. Some of these toilets even clean themselves once we've left, applying an antibacterial coating as the last step.

But while smart toilets currently exist—even if they haven't reached our shores yet—even smarter ones are in development. Their purpose: to keep us healthy. Some toilets in Japan already perform urinalysis to see if users have diabetes; soon there will be toilets able to detect things like drug use and pregnancy from your urine, plus colon cancer from your stools. Heck, they'll even be able to give us diet and exercise advice. But will we heed it?

5. SCREENING PROCESS: HOME SECURITY THROUGH FACIAL RECOGNITION SOFTWARE

Lots of movies today feature businesses and government entities with facial recognition software in place, usually to keep unauthorized personnel from accessing areas with highly classified information and priceless goods. Eventually, many of us may be able to install this software in our own homes.

Why would we want to? Video cameras at our doors could identify our family and friends, plus strangers. If a stranger does ring our doorbell, the system could then immediately run the person's mug against all the faces in criminal and terrorist databases, so we don't let in some really bad dude. Of course, sometimes a family member or friend turns out to be a bad guy.

There are numerous companies currently working with facial recognition software for both business and personal use. When and if it becomes widely available isn't yet known. Android currently has a face-unlock feature in its Android 4 operating system, aka Ice Cream Sandwich. But if you hold up a photo of the authorized user, it can be fooled. So, it seems we have a way to go on this one.

INTELLIGENT TRANSPORTATION SYSTEMS: THE FUTURE OF TRAVEL?

Rapidly advancing technology during the past few decades has changed how we work, how we entertain ourselves, and how we connect with one another in our homes and other buildings. But what about how we get to and from those places and elsewhere? Now advances in technology promise to improve how we drive.

According to an IBM white paper titled "The Case for Smarter Transportation," in 2007 Americans "wasted 4.2 billion hours, 2.8 billion gallons of fuel, and $87.2 billion due to traffic congestion."

Intelligent transportation systems are a vision of a future that integrates existing transportation infrastructure with communication networks in an effort to reduce congestion and travel time. In doing so on a mass scale, the larger effect of intelligent transportation systems is to limit the release of carbon emissions into the atmosphere, cut back on fuel consumption, and improve road safety.

It's hard to argue that technology hasn't already had an impact. Onboard computers maximize engine performance and lead to a safer ride. Hybrid cars have created a new class of vehicle with higher fuel efficiency. GPS systems ensure that passengers get to their destination as efficiently as possible.

But intelligent transportation systems imply improvements not only in vehicle technology but also in the creation of an integrated network linking cars and trucks with roadway infrastructure. Because many of these component technologies are still in the theoretical or experimental phase, there are a range of prescriptions for setting a standard for deploying technological improvements on our nation's transportation infrastructure.

Although specific applications require research, testing, and pilot studies before they could be deployed on a scale large enough to have an impact on urban traffic congestion, different approaches share a combination of high- and low-tech solutions to traffic problems.

WEATHER WARNINGS: HOW INTELLIGENT CARS WILL CUT CONGESTION

Severe weather, road hazards, and accidents can add considerably to the travel times of every driver passing along the same route. Although many newer GPS devices are equipped with traffic information, these devices simply aren't ubiquitous—or often accurate—enough to make a significant dent in congestion created by these kinds of incidents.

Cell phones, computers, and tablet devices are already capable of sending and receiving data. So why not cars as well?

Individual cars could essentially act as data points on a network. These kinds of networks could have immediate benefits to drivers, who would be quickly rerouted in the event of a congestion-inducing incident before traffic can build to the point of adding significant delays. These data can also help consumers decide whether driving really is the best option on a given day or if public transportation offers a more efficient means of conveyance.

In the long term, daily commuter information, traffic patterns, and incident reports can be used to help transportation officials and city planners determine future roadwork and safety projects.

IN THE MEANTIME...LOW-TECH TRAFFIC SOLUTIONS

Since people cannot expect to see a fully realized and integrated transportation communications network arriving in their neighborhoods anytime soon, iterative changes to existing roadways using relatively low-tech alternatives could provide an interim solution.

Responsive traffic lights are increasingly common on roadways across the country. Variable speed limits,

in which highways permit higher speeds during times
when transportation authorities anticipate less traffic,
can be another simple improvement for drivers.

Up to $30 trillion will be spent on improvements to
U.S. infrastructure in the next twenty years, according to the
same IBM white paper cited previously. So although these
roadways of the future will take a significant investment to
get off the ground, drivers should begin seeing the benefits
coming around the corner soon enough.

FIVE OF THE COOLEST CAR TECHNOLOGIES THAT TRULY HAVE A CHANCE

n the technology world, the latest advancement is only as good as the next thing coming down the line. The auto industry is constantly bringing us new technologies, whether for safety, entertainment, usefulness, or simply for pure innovation.

Many new car technologies are either specifically built for safety or at least have some sort of safety focus to them. Some of the latest car innovations we've found are truly exciting technologies that could revolutionize not just the automotive industry but also human transportation in general.

So what's in store for future cars? Well, we don't know for sure, but based on what's currently being tested and what's on the road today, we have an idea of some new technology that will most likely make it into production. Some of it will help keep us safe, some will give us information like never before, and some will let us kick back and just enjoy the ride. Here are five of the coolest upcoming car technologies.

ⅢⅢ 1. CARS THAT COMMUNICATE WITH EACH OTHER AND THE ROAD

We looked at this a little bit in the previous section, but car manufacturers and the U.S. government are seriously looking into and researching two technologies that would enable future cars to communicate with each other and with objects around them.

Imagine approaching an intersection as another car runs a red light. You don't see the car at first, but your car gets a signal from the other car that it's directly in your path and warns you of the potential collision, or even hits the brakes automatically to avoid an accident. A developing technology called vehicle-to-vehicle communication (V2V) is being tested by automotive manufacturers like Ford as a way to help reduce the number of accidents on the road.

V2V works by using wireless signals to send information about each car's location, speed, and direction back and forth between cars. The information from one car is communicated to the cars around it in order to provide information on how to keep the vehicles safe distances from each other.

At the Massachusetts Institute of Technology, engineers are working on V2V algorithms that calculate information from cars to determine what the best evasive measure should be if another car started coming into a car's own projected path. The National Highway Traffic

Safety Administration (NHTSA), in a 2010 report, stated that V2V has the potential to reduce 79 percent of target vehicle crashes on the road.

But researchers aren't only considering V2V communication. Vehicle-to-infrastructure communication (V2I) is being tested as well. V2I would allow vehicles to communicate with things like road signs or traffic signals, which could provide information to the vehicle about safety issues. V2I could also request traffic information from a traffic management system to access the best possible routes. Reports by the NHTSA state that incorporating V2I into vehicles, along with V2V systems, would reduce all target vehicle crashes by 81 percent.

These technologies could transform the way we drive and increase automotive safety dramatically. Good thing car companies and the government are already working to make this a reality. All of this communication and preemptive vehicle assistance leads us into our next future technology.

2. SELF-DRIVING CARS

The idea of a self-driving car isn't new. Many TV shows and movies have had the idea, and there are already cars on the road that can park themselves. But a real self-driving car means exactly that—one that can drive itself—and it's probably closer to being a reality than you might think.

In California and Nevada, Google engineers have already tested self-driving cars on more than 500,000 miles (804,672 kilometers) of public highways and roads. Google's cars not only record images of the road, but their computerized maps also view road signs, find alternative routes, and see traffic lights before they're even visible to a person. By using lasers, radar, and cameras, the cars can analyze and process information about their surroundings faster than a human can.

If self-driving cars do make it to mass production, we might have a little more time on our hands. Americans spend an average of thirty-eight hours sitting in traffic every year. Cars that drive themselves would most likely have the option to engage in platooning, where multiple cars drive very closely to each other and act as one unit. Some people believe platooning would decrease highway accidents because the cars would be communicating and reacting to each other simultaneously, without the ongoing distractions that drivers face.

In some of Google's tests, the cars learned the details of a road by driving on it several times, and when it was time for a car to drive itself, it was able to identify when there were pedestrians crossing and stopped to let them pass by. Self-driving cars could make transportation safer for all of us by eliminating the cause of 95 percent of today's accidents: human error.

Although self-driving cars may seem far off, GM has

already done its own testing, and some people believe that you'll see some sort of self-driving car in showrooms in the next decade.

Read on to learn how we may be viewing all of our car's data in the near future.

||||| 3. AUGMENTED REALITY DASHBOARDS

GPS and other in-car displays are great for getting us from point A to point B, and some high-end vehicles even have displays on the windshield, but in the near future, cars will be able to identify external objects in front of the driver and display information about them on the windshield.

Think of *The Terminator* or many other science fiction stories where a robot looks at a person or an object, automatically brings up information about them, and can identify who or what they are. Augmented reality (AR) dashboards will function in a similar way for drivers. BMW has already implemented a windshield display in some of its vehicles that shows basic information, but it's also developing AR dashboards that will be able to identify objects in front of a vehicle and tell the driver how far that person is from the object. The AR display will overlay information on top of what a driver is seeing in real life.

So if you're approaching a car too quickly, a red box may appear on the car you're approaching and arrows will appear showing you how to maneuver into the next lane before you collide with the other car. An AR GPS system could highlight the actual lane you need to be in and show you where you need to turn down the road, without you ever having to take your eyes off the road.

NOT JUST FOR DRIVERS

BMW is also researching the use of AR for automotive technicians. The company produced a video where a BMW technician uses AR glasses to look at an engine and identify what parts need to be replaced, and then it shows step-by-step instructions on how to fix the car.

AR is being researched for passengers as well. Toyota has produced working concepts of its AR system that would allow passengers to zoom in on objects outside the car and select and identify them, as well as view the distance of an object from the car using a touch-screen window.

AR may not be here yet, but if these car companies have their way, we'll be seeing it in our future cars down the road.

▥ 4. AIRBAGS THAT HELP STOP CARS

Ever since airbags were added to vehicles, they've continued to make their way around the inside of our vehicles. We now have curtain airbags, side airbags, knee airbags, seat-belt airbags, and even ones that deploy under us. Maybe all of us don't have them in our cars, but they're on the road.

Mercedes is experimenting with a new way to use airbags that deploy from underneath the car to help stop a vehicle before a crash. These airbags are part of the overall active safety system and deploy when sensors determine that an impact is inevitable. The bags have a friction coating that helps slow the car down and can double the stopping power of the vehicle. The bags also lift the vehicle up to three inches (eight centimeters), which counters the car's dipping motion during hard braking, improves bumper-to-bumper contact, and helps prevent passengers from sliding under seat belts during a collision.

What gives this kind of airbag potential as a future technology is that it uses existing vehicle safety systems. Although Mercedes has been working on this technology for several years, it isn't available on any production models yet and may not be seen on the road for another few years.

With the current evolution of airbags and their pervasiveness within the automotive world, it wouldn't be a stretch to imagine future cars using airbags to not only protect passengers, but also stop cars.

ⅢⅢ 5. ENERGY-STORING BODY PANELS

ExxonMobil predicts that by 2040, half of all new cars coming off the production line will be hybrids. That's great news for the environment, but one of the problems with hybrids is that the batteries take up a lot of space and are very heavy. Even with advances in lithium-ion batteries, hybrids have a significant amount of weight from their batteries. That's where energy-storing body panels come in.

In Europe, a group of nine auto manufacturers are currently researching and testing body panels that can store energy and charge faster than today's conventional batteries. The body panels being tested are made of polymer fiber and carbon resin that are strong enough to be used in vehicles and pliable enough to be molded into panels. These panels could reduce a car's weight by up to 15 percent.

The panels would capture energy produced by technologies, like regenerative braking or energy gathered when the car is plugged in overnight, and then feed that energy back to the car when it's needed. Not only would this help reduce the size of hybrid batteries, but the extra savings in weight would also eliminate wasted energy used to move the weight from the batteries. Toyota is also looking into energy-storing panels, but they're taking it one step further and researching body panels that would actually capture *solar* energy and store it in a lightweight panel.

Whether future body panels collect energy or just store it, automotive companies are looking into new ways to make our cars more energy efficient and lightweight.

SMART STREETS WITH STREET SMARTS: HOW INTELLIGENT HIGHWAYS WILL WORK

When you are on the road during the holidays, you probably spend some time staring at the bumper in front of you. Can you imagine a world without gridlock?

The city of Toronto sure could use that. The main artery for traveling in and out of Toronto, Ontario, is Highway 401, a thoroughfare that expands up to fourteen lanes at its widest. And with more than 350,000 vehicles per day, including 45,000 trucks, Highway 401 is only exceeded in traffic volume by the Santa Monica Freeway in Los Angeles. "It's world-class congestion. It comes to a grinding halt at rush hour virtually every day," said Brian Marshall of the Canada Transportation Development Centre.

Traffic is a growing problem in almost every city in the world. The cost of traffic congestion in the United States alone is $78 billion, representing the 4.5 billion hours of travel time and 6.8 billion gallons of fuel wasted sitting in traffic. Billions more dollars have been spent on electronics and systems to alleviate this logjam.

Government transportation agencies are seeking out

new, cheaper technology to replace the high-priced loop sensors and other invasive technologies that have been used in the past. In this section, we will drive on the freeway of the future and see how ubiquitous digital devices will aid in easing our traffic woes.

‖‖‖‖ CURRENT TRAFFIC TRACKING

The next time you are driving to work, take a minute to look at the technology that keeps traffic flowing. Over the past two decades, state departments of transportation have installed billions of dollars' worth of electronics, like loop detectors, video cameras, and electronic display signs, to keep an eye on and manage traffic.

Loop detectors are wires embedded in the road that detect small changes in electrical voltage caused by a passing vehicle. Traffic speed can be determined by detecting how quickly cars pass between two sets of loop detectors. Volume and speed data is transmitted to a central computer, which is monitored by local transportation departments.

If the detectors sense a slowdown or an increased quantity in traffic, workers can use video cameras to get a better understanding of what's causing it. Meanwhile, messages can be displayed on electronic signs to warn motorists of congestion ahead and to advise alternate routes.

The problem with the system, Marshall said, is that "the

traditional loops in the road and cameras up on poles and guys sitting behind desks looking at monitors are too expensive to extend as far as people would like." Installing these detectors, cameras, and signs has been a long process and is costing billions of dollars for state and federal governments. Transportation officials are now searching for cheaper alternatives for managing traffic.

TAGGING TRAFFIC

You are in a shrinking minority of the American population if you don't own a cell phone. Some 90 percent of U.S. adults own a cell phone of some kind as of January 2014. Each day, thousands more people sign up for mobile services, and smartphones are becoming increasingly popular. As a result, cell phone technology, particularly devices with GPS capabilities, are changing the way that we navigate. As another example, transportation agencies are installing additional electronic toll-tag readers along major highways. In some cities where tollbooths are common, radio-frequency tags are attached to cars. As a car passes, the reader detects the tag and subtracts a set amount of money from a prepaid account.

These radio tags, or transponders, can be used to time vehicles between points in a freeway system. Unlike with a tollbooth, drivers do not have to slow down for the reading device. They merely drive past it. By analyzing a particular car's time between two points, a computer can determine the car's location and speed.

These tags and GPS tracking systems will make it almost impossible for someone to travel undetected, which has raised privacy concerns about this technology. For instance, some companies have said that they are considering selling customer information to advertisers.

CAUTION: ACCIDENT AHEAD

Once information is detected from cell phones, it has to be disseminated to motorists. In order for drivers to be routed around traffic, they must be informed of how fast the traffic is flowing, if it's clogged, or if there is an incident blocking traffic altogether. This is where a cell phone service provider comes into the picture. The provider would send this information out to customers.

There are three ways to transmit information to motorists:

- Collected information is fed into a large repository that can be accessed via a website or app. A map on the screen would show various roadways in green, yellow, and red to indicate free-flowing traffic, slow traffic, and clogged traffic, respectively. Google Maps, for example, provides much of this functionality for six hundred areas in more than fifty countries.

- Registered users, whose locations are known, are sent customized traffic reports based on the road and direction in which they are traveling. Systems will also advise users of alternate routes around congested areas.
- Information is displayed on conventional electronic road signs.

By getting information to drivers quicker, developers believe that commuters will have enough time to react to these warnings and find another way around the congested areas. This would be an improvement over how information is released today, which is primarily through radio or television news reports. By the time the radio and TV report an incident, it's typically too late for most commuters to act on the information.

Cell phones and other digital devices are as commonplace as cars, so why not combine the two to solve the problem of congested highways? In the next few years, we will see whether these new technologies will make our commute to work easier or if our only hope is to find a way to stay home.

CHARIOTS OF HIGHER: HOW SOLAR AIRCRAFT WORK

Now that we've covered the roads, let's take to the skies. We can make our automobiles green, but what about our airplanes? The answer is yes, and the technology is already out there in solar aircraft.

Although they've been flying since the 1970s, solar aircraft may have flown so far below your radar that they sound new. A solar airplane could take you for quite an amazing ride. You'd have to start in the morning and wait for the clouds to clear. Propellers whirring, the plane would travel with yawn-inducing slowness down the runway. As the wind caught the plane, you'd ascend so slowly that you'd hardly be pressed into your seat.

You'd climb above birds, above Mount Everest, above commercial jets, and above military spy planes. (NASA and AeroVironment's Helios climbed to 96,863 feet, or 29,524 meters.) You'd settle into the stratosphere, home to icy cirrus clouds.

However fun such a joyride might sound, solar planes are designed for other uses. Since they're basically low-flying satellites, at first NASA envisioned parking them over cities as communications platforms, but that was before we had so many cell phone towers. Now, the military is eyeing solar planes for surveillance.

They can, in theory, stay aloft for years. In reality, though, the stats aren't quite there yet. The record is two weeks (really!) without landing, set by QinetiQ's unmanned Zephyr. For manned aircraft, the record is twenty-six hours, ten minutes, and nineteen seconds, held by the Swiss aircraft Solar Impulse. With records like these on their side, some organizations hope to change attitudes that solar power is weak and inefficient.

Probably the easiest way to understand how solar aircraft work is by comparing them to more common airplanes in the sky. We'll look at one commercial jet—Boeing's 747-400—and one military jet—the F-22A Raptor.

SOLAR VERSUS TRADITIONAL AIRPLANES

As mentioned earlier, solar airplanes are mostly surveillance craft. Boeing's 747-400, on the other hand, flies from Detroit to Tokyo, carrying hundreds of passengers on decent fuel mileage. The F-22A Raptor, by contrast, is a fighter plane for the U.S. Air Force. It's designed to be fast, agile, quiet, and almost invisible. These are the basic differences. Let's put these planes head to head, or wing to wing, to find out even more.

SPADES AND FLYING RULERS

Many solar planes are shaped like flying rulers. NASA's Helios plane, for instance, had a 247-foot (75-meter) wingspan but was only 12 feet (3.7 meters) long. A 747's wingspan is shorter, at 211 feet (64.3 meters), and its fuselage is about the length of its wingspan. The F-22A Raptor is a spade-shaped, stubby plane, 44.5 feet (13.6 meters) across the wings and 62 feet (18.9 meters) long.

Compared to the other planes, solar planes are practically kites!

- Some are launched by hand with a running toss into the air.
- The Helios, which was lost in a crash in 2003, was too heavy for that. It weighed 2,048 pounds (929 kilograms) at most, made of pricey, light, and strong materials—and Styrofoam. Amazingly, the whole plane bent. (More on that later.)

Traditional planes are veritably obese compared to solar ones:

- A Raptor weighs a formidable 83,500 pounds (37.875 kilograms) and is most definitely not bendable.
- The 747 weighs up to 875,000 pounds (396,893 kilograms), including all the luggage in the cargo hold.

You'll find a lot of electric propellers—up to fourteen—on a solar airplane, and that's all of its propulsion. Of course, electric propellers wilt next to jet engines. The Raptor's jet engines shoot it forward with 70,000 pounds (311,500 newtons) of total thrust. A 747's two engines move it with up to 126,600 pounds of total thrust (563,145 newtons).

FROM ELECTRIC CARS TO ELECTRIC AIRPLANES!

Solar power may not be the only way to help planes efficiently take to the skies. Electric aircraft may not be all that far off in the future—and they would be a very fuel-efficient and environmentally friendly way to fly.

You wouldn't be surprised by which one would win in a race. While the environment smiles on pollution-free solar planes, the gods of speed do not. When cruising at low altitudes, the Helios traveled no more than 27 miles per hour (43.5 kilometers per hour). A 747 cruises at 567 miles per hour (913 kilometers per hour), and the Raptor can reach close to Mach 2.

The Raptor also wins on maneuverability. While the 747 can turn, pitch, and change its speed, as could the Helios, the Raptor can fly a spinning loop-the-loop.

So far, the 747 wins on distance. The farthest flight for a solar plane has been 693 miles (1,115 kilometers). The Raptor's maximum range is 1,841 miles (2,963 kilometers), while the 747 can fly 8,355 miles (13,446 kilometers).

Solar planes win in a category you probably haven't considered—longevity. Jets must land to refuel. Solar planes don't have to. They can stay aloft as long as their batteries are charged to get them through the night. By staying aloft for more than three days, solar planes have already surpassed jets, and many solar-plane makers share the goal of extending that to months or years.

Now that you know how a solar plane stacks up against other aircraft, let's take a closer look at its design.

LIMITED BY THE SUN

The solar energy that hits a square foot of panels on a solar plane in an hour is tiny compared to the energy in a gallon of jet fuel. Solar panels also convert less solar power—between 10 and 20 percent—to electrical power for turning the propellers, compared to the amount of power combustion in a jet engine applied to thrusting a jet forward. In the end, 1 square foot (0.09 square meter) of solar panels yields three to six times less power than you need to light a 60-watt light bulb, so you can see why engineers paper the plane with panels and try to keep them light.

SOLAR AIRCRAFT DESIGN

Solar airplanes don't have much on board. They're ridiculously flat and thin, inviting the wind to lift them instead of knocking them around. The body is strong and light, often made of carbon-fiber pipes for the frame, with a strong fabric like Kevlar stretched across it. Somewhere in the structure, you'll see an X or V, which prevents the plane from rolling.

Most solar planes run on batteries at night, although some have used fuel cells. The batteries are light and energetic and are usually arranged in a sheet. QinetiQ's solar plane Zephyr, which holds the current endurance record, runs on a sheet of lithium-sulfur batteries. The batteries are wired to motors that turn propellers.

You can't miss the solar panels, which are the skin and heart of the plane. They don't resemble the rigid, bulky solar panels on satellites or a solar house. The solar plane's panels are millimeters thick, flexible enough to roll, and incredibly efficient and expensive. The solar panels are also wired to the propellers.

On board, the plane will carry light, voltage, and wind sensors, and it will have a method for relaying that information to the pilot.

If you're wondering where the wheels are, there's no need to bother. As John Del Frate, an engineer at NASA's Dryden Flight Research Center, explains: "Some solar airplanes basically drop off the landing gear in flight because you're not going to need them. The plane may land on skids or crash-land. Engineers are getting rid of every bit of weight you can possibly imagine."

Some solar airplanes are true unmanned aerial vehicles (UAVs). Except for takeoff and landing, an autopilot flies the plane. Pilots use onboard systems to track the plane and control its motors from the ground. Unmanned planes include NASA's deceased Helios, the Zephyr, and Aurora Flight Sciences's Odysseus and SunLight Eagle.

Other solar planes can support a pilot. Examples of piloted solar aircraft are NASA and AeroVironment's retired Gossamer Penguin and Solar Challenger, and the Solar Impulse, developed in Switzerland at the École Polytechnique Fédérale de Lausanne, which aims to circumnavigate the globe.

STRANGE SOLAR

The Odysseus, concocted by Aurora Flight Sciences, is still in design. In simulations, three segments of plane take off from the runway. With the segments joining together, the plane builds itself midair. To catch the sun from more angles, the plane folds itself like an accordion.

You might be skeptical, but DARPA is not. It selected the Odysseus for its Vulture Program, which will try to fly a plane continuously for five years while carrying 1,000 pounds (454 kilograms).

IIIII FLYING SOLAR

A solar plane's flight starts with checks. Check the battery—it should be charged. Check the ground winds—they shouldn't exceed about 10 miles per hour (16 kilometers per hour), or else the plane could crash on the runway. Check for turbulence in the air because the plane will have to ascend through the turbulent layers. Billows in the clouds are a bad sign. "Wind is your enemy," says Del Frate.

Morning is best for takeoff, when the sun is overhead and there are ample hours of sunlight left in the day. As a runway, you'll need a circle a little more than three football fields across, which is ten times shorter than an average airport runway. Next, you angle the plane for takeoff, using that circle. You point the plane so the wind blows head-on, but never across it. Crosswinds spell destruction for most solar planes because they can throw the plane in unwanted directions.

When the propellers are online, a combination of battery power and solar energy can start them spinning, and the plane is ready to roll (or be hand-tossed into the air). "The plane takes off at bicycle speeds," says Del Frate, because takeoff is typically done on solar power. As a pilot, usually on the ground, you avoid shadows and steer for maximum sun to preserve battery power.

The plane ascends slowly. You make it ascend by speeding its central propellers, tilting it up. By 35,000 feet (10.6 kilometers) or so, you've hit the jet stream. Hold on. In this turbulent layer of the sky, planes can bend, like NASA's Helios did, from flat to a dramatic U, with the wind. If the plane didn't bend, the wind could rip it apart. The plane can't stay here, where 747s cruise.

Above the jet stream, dodge the puffy clouds—they block the sun. Turning is as easy as speeding the propellers on one side of the airplane. By 40,000 feet (12 kilometers), you've entered the stratosphere, a still layer with icy cirrus clouds that don't block the sun. Finally, by 65,000 feet (20 kilometers), you can relax in stillness and practically glide. If you plan to stay up overnight, make sure your battery is charged to run the propellers. Otherwise, you'll start losing altitude.

During flight, a solar plane switches automatically between battery and solar power. When there's sun, it runs the propellers and charges the batteries or fuel cells. To charge the battery faster, the pilot can fly slower. At night or in clouds, the propellers run on the battery or fuel cells alone.

When it's time to land, cut the power to stop the propellers. Solar planes glide down—engineers would rather make them efficient fliers than fast at landing. "They descend extremely slowly," says Del Frate. "When you're trying to bring one in for a landing, you'd like to grab it and pull it down."

SOLAR SMARTS: ENVIRONMENTAL BENEFITS OF SOLAR AIRCRAFT

Many researchers say it's useful to park a solar aircraft in the sky. It can hover over a spot, carrying cameras or other sensors. In the stratosphere, it can sample gases near the ozone layer. It can also watch forest fires or track hurricanes on the ground.

For the military, solar airplanes can help with

reconnaissance. Like spy planes, they fly high, which makes them stealthy. But while spy planes must fly over enemy territory and return, solar airplanes are unblinking eyes. They can take uninterrupted photos or videos for years. "When an event happens, they can study everything that led up to it," says Del Frate. For law enforcement, they're good for border and port patrol.

It's true that satellites can perform some of these tasks. But solar airplanes see more detail on the ground with less expensive cameras because they're closer to the action. They're also less expensive to build and launch. While satellites are hard to move once they're in orbit, solar airplanes are easily moved. It's also easier to bring solar planes down for maintenance.

Solar aircraft, being electric, emit no exhaust. Commercial airplanes do. In 1992, airplanes emitted a half billion tons of CO_2, or 2 percent of human CO_2 emissions alone! Since then, the numbers have only risen. Airplane exhaust contains many substances linked to health and environmental effects, although the U.S. Environmental Protection Agency (EPA) regulates their levels, and health impacts near airports are being studied. Regardless, solar planes can't become clean passenger planes because they'll probably never have enough power to carry many passengers, according to Del Frate.

Stratospheric jets, like the F-22A Raptor and U-2 spy planes, also emit exhaust. While they emit it into the stratosphere, where gases persist longer than in our troposphere below, their contribution to air pollution, ozone depletion, and global warming hasn't been measured thoroughly. Solar airplanes that can accelerate and maneuver like these planes are many years off. So at this time, it's not practical to talk about solar planes being environmentally friendly alternatives to other planes. Still, they are clean vehicles for their current applications.

A surprising benefit of solar airplanes, explains Del Frate, is that if solar panel manufacturers supplied a dozen solar planes a year with big, high-efficiency panels, the cost of high-efficiency panels for your home would go down.

IT'S A BIRD! IT'S A PLANE!

"Someone once asked me, what about birds? Are you worried about hitting a bird?" recalls NASA engineer John Del Frate. "I had to laugh. A bird can outfly one of these things. A bird could make a nest on one of these things while it's flying!" Because it's hard to move a solar plane out of the way, pilots work with the Federal Aviation Administration to plan flights away from other planes. Luckily, very few planes fly at 65,000 feet (20 kilometers).

IIII SOLAR SCARES: CONCERNS ABOUT SOLAR AIRCRAFT

"I remember that people never thought they'd be able to fly. After the planes set flight records, those critics were silenced," says Del Frate. But critics still take issue with solar airplanes.

"Critics tend to point out that these airplanes are fragile," says Del Frate. NASA's Pathfinder plane was damaged inside a NASA hangar by wind blowing through the door. "We build in the necessary strength and no more. They're light and very minimal on material—for a reason."

It's hard to imagine paying $20 million for a plane as thin as a wafer, but that's about what solar airplanes cost. According to Del Frate, the solar panels alone account for about half the cost. But to put it in perspective, a Boeing 747 starts at $234 million.

The planes are not heavy lifters; the strongest built to date can carry one pilot. "If they hardly carry any payload, what's the point?" says Del Frate, summarizing what critics say. He points out that solar planes can carry sensors and cameras, which are light and getting lighter. "Look at all your cell phone can do. It hardly weighs anything."

So far, solar planes need special flight conditions. While the batteries can carry them through night and the shade, the planes can't take off or fly in storms. They can't take off in strong wind. They can't stay in cumulus clouds or turbulent layers of the sky.

"Critics will point out they're only useful nine months out of the year, and they're right," says Del Frate. During winter, the planes struggle to stay up, with days being short and nights being long. Because the sun is close to the horizon and the solar panels usually point straight up, the plane struggles to collect enough sunlight to stay aloft. Designers are angling and placing solar panels to catch the sun no matter where it is—and some are experimenting with folding planes.

HELIOS TOTALED BY TURBULENCE

Ironically, turbulence—the very thing Helios was designed to avoid—probably took down the NASA aircraft during a test in Hawaii in 2003. It all started when a swirling wind at 2,800 feet (853 meters) bent the wings up, which tilted the plane down. "It started to oscillate in pitch—nose up, nose down, nose up, nose down—and each time it oscillated, it doubled its speed. When it was going three times faster than it was designed to fly, the solar cells began to strip off. Pretty soon, we lost all lift, and it fell into the Pacific Ocean," says Del Frate.

4

FIXTURES IN THE FUTURE

So now that we've looked at homes and hot new ways to travel, let's take a look at future fashion fixes and some final future trends and tall tales!

FABRIC, FABRIC, EVERYWHERE! HOW FABRIC DISPLAYS WORK

Some people are just begging for attention. Marketers are constantly trying to find ways to build brand awareness, often with clothing—it's a common practice to make shirts and hats featuring company logos and slogans. To really grab your attention, some companies are using fabric displays—techniques and systems designed to make dynamic images and text on clothes and other things made of fabric.

There are many different kinds of fabric displays. Some use a still image as a starting point, relying on fabric with special properties to make the design more eye-catching. Other fabric displays can show full video with sound. Each method relies on different technologies, and all have their advantages and disadvantages.

WATCH YOUR SHIRT

Fabric displays are revolutionizing the marketing and fashion industries. From furry TVs to wearable computer systems, these displays are worth watching.

A few fabric display techniques are readily available to the consumer market. Creative individuals have used fabric display technology to build elaborate costumes. Jay Maynard used electroluminescent (EL) wire in the costume he built based on the Disney film *Tron*. His efforts gained national attention, and before long Maynard was making the talk-show circuit as the "*Tron* guy."

In this section, we'll look at the different ways inventors have modified clothing to make a bigger impact on audiences. We'll learn about an idea for fur displays that uses electrostatic charges to shocking effect. We'll see how a heat-sensitive dye can turn a normal T-shirt into a very large mood ring.

After that, we'll explore the world of EL clothing. Then we'll see how LED and PLED displays can turn a normal outfit into an eye-catching light display. Finally, we'll learn about companies that have created clothing with built-in television and PC displays.

FUZZY WUZZY WAS A FABRIC DISPLAY

Up first: fur. There has been some confusion about what, exactly, a fur fabric display is. Philips Electronics filed a patent application with the simple title "Fabric Display," though some science blogs and magazines have referred to it as "furry television." At its most basic level, this fur fabric display relies on a very simple technology. Patches of fur cover an image, and when the fur moves, it reveals the image underneath. It's a simple way to conceal and reveal designs.

The fabric display has three layers. The bottom layer is conductive, which means it can carry electricity from a power source—like a small battery pack—to the rest of the fabric to create an electrostatic field across the fur, which gives each strand of fur the same electrical charge.

The next layer in a fur fabric display is the fabric's base color or design. This could be a company logo, a picture, or just a particular color. The furry display doesn't change the design on the cloth; it just hides or reveals portions of the design at a given time.

The third layer is the fur. It can be any color, but it must be short enough so that when the user turns on the electrostatic field, the strands stand on end and reveal the design or color of the fabric underneath. For example, in a simple fur fabric display, you could use red fur to cover a blue shirt. When you turn on the power for the conductive layer, the red fur would stand on end, revealing the blue shirt underneath. To a distant observer, it would appear that the shirt had just magically changed colors.

The patent application refers to each small, visible section of the base fabric as a "pixel," which may be why some articles refer to the display as furry television. While it might be possible to approximate primitive animation techniques by printing one image across the fur layer and a slightly adjusted image on the fabric underneath, it's not quite the same as watching television on someone's jacket.

In the next section, we'll learn how some designers use heat to create fabric displays.

FUR FABRIC FUNCTIONALITY

To understand static electricity, we need to start all the way down at the atomic level. All matter is made up of atoms, which contain charged particles. Atoms have a nucleus consisting of neutrons and protons and a surrounding shell that is made up of electrons. Neutrons carry no charge, but protons have a positive charge and electrons have a negative charge. So, if an atom has more protons than electrons, its charge is positive, and if it has more electrons than protons, its charge is negative. Similar charges repel each other, and opposite charges attract each other.

An electrostatic field applied to fur creates a similar charge across the fur and the base of the material. Since similar charges repel one another, the fur moves as far away from the base of the material and other strands of fur as it possibly can, causing the strands to stand on end.

IT'S GETTING HOT IN HERE: THERMOCHROMIC FABRIC DISPLAYS

The word "thermochromic" looks a little intimidating at first, but the concept itself is pretty simple. *Thermo* comes from the Greek word "thermos," which means warm or hot. *Chromic* comes from "chroma," meaning color. A thermochromic substance changes color as it changes temperature. In fabrics, a special dye acts as the thermochromic agent.

Some thermochromic dyes change from colorful to clear, revealing the color of the fabric underneath. Companies can use thermochromic dyes in shirts that slowly reveal a company slogan or logo as the shirt heats up. When the shirt cools down, the logo seems to disappear.

There are two widely used elements in thermochromic dyes, and both rely on chemical reactions:

- **Liquid crystals:** These thermochromic dyes rely on liquid crystals contained in tiny capsules. The liquid crystals are cholesteric, also known as chiral nematics, which means that the molecules arrange themselves in a very specific helical structure. These structures reflect certain wavelengths of light. As the liquid crystals heat up, the orientation of the helices changes, which causes the helices to reflect a different wavelength of light. To our eyes, the result is a change in color. As the crystals cool down, they reorient themselves into their initial arrangements and the original color returns.

- **Microencapsulate thermochromic system:** In this system, the thermochromic dye contains millions of tiny capsules that each look a little like an organic cell. Each capsule has an outer membrane and contains an organic, hydrophobic solvent, which makes it less likely that water will dilute or wash out the chemicals in the dye. The solvent contains particles of a color developer and a dye precursor. As the capsule heats up, the solvent melts and a chemical reaction causes the color developer to donate a proton to the dye precursor. In turn, this causes the precursor to develop into the dye itself and change color. When the dye cools down, the developer and precursor separate, the solvent resolidifies, and the color returns to its original state.

Like fur fabric displays, thermochromic fabrics aren't animated—they can only conceal and reveal designs or colors based on environmental conditions. While that might be enough for some people, others want even more dynamic clothing.

In the next section, we'll look at a technology that turns normal clothes into wearable neon signs—EL fabric displays.

BLAST FROM THE PAST

In the late 1980s, a company called Generra introduced a line of clothing that used thermochromic
dyes. The company called the line Hypercolor. Schools
across the United States were filled with students wearing the popular shirts and hats, most of them bearing
a collection of handprints where other kids left their
temporary marks.

TIE-DYE NO MORE: EL FABRIC DISPLAYS

If wearing a furry display or heat-sensitive clothing doesn't
seem appealing, you might want to look into EL fabric displays. EL substances give off light after being exposed to
electricity. For fabric displays, designers use EL wire to
create amazing, vibrant effects.

EL wires have several layers:

- The core layer is a copper wire that acts as a conductor
  in the EL wire's AC power system.
- On top of the copper is a coating of EL phosphor. This
  is the material that will emit light after encountering an
  AC electric field.
- The next layer consists of two wires wrapped around

the phosphor. These wires complete the second half of a circuit, the first half consisting of the copper conductor.

- Last comes a pair of plastic sheaths, which protects both the phosphor material (moisture can ruin some phosphors) and the user from electric shocks.

EL wire needs a high voltage—around 100 volts—to glow brightly. Lower voltages result only in a dull glow. Some EL wires can produce a range of light wavelengths depending on the frequency of applied power. Also, because EL wire needs an AC power system, any outfit that has EL wire will need a battery pack and an inverter—a device that converts DC power to AC power.

Because the core of EL wire is copper, it's flexible but holds its shape. You can bend EL wire into all sorts of designs. When it's turned off, EL wire looks like a colorful plastic tube, but when the power comes on, EL wire looks like thin strands of neon lights. An outfit with EL wire could have several different strands emitting different colors, and might even include a sequencer—a special circuit board—connected to the power source that manages each strand's power supply. By alternating power to various strands, the wire can appear to be animated as different strands flash on and off.

Clothes with EL wire require careful maintenance and

cleaning procedures. If the wire is permanently affixed to the clothing, the wearer will need to carefully wash it by hand and let it dry on a flat surface, or rely on spot cleaning. Throwing EL clothes into the washing machine is a good way to ruin a special outfit, and could even damage other clothes or the washing machine itself if the plastic tubing around the copper core tears.

EL clothes are bright and vibrant, and with the right equipment they can display lights in patterns and sequences, but they're still fairly static—you're limited by the shapes into which you've bent the EL wire.

In the next section, we'll learn about LED fabric displays and how they give you more options to express yourself through clothing.

PHOSPHORS

EL substances are a kind of phosphor. Phosphors are materials that emit light after absorbing some form of radiation or as a result of a chemical reaction. EL phosphors emit light after absorbing electricity.

ⅢⅢ THIS LITTLE LIGHT OF MINE: FABRIC DISPLAYS USING LEDs

Light emitting diodes (LEDs) are tiny light bulbs designed to fit into electrical circuits. Unlike incandescent bulbs, LEDs don't use a filament to generate light. Instead, light is a by-product of electron motion within semiconductor material. Electrons move from high energy states to lower ones, releasing photons in the process. LEDs take advantage of this, harnessing and focusing the photons into tiny light bulbs. The gap between the higher and lower states of energy determines the frequency of the photon, which we observe as a specific color of light.

Several companies sell clothes that use LEDs to create special patterns or messages. Most of these companies will alter regular clothing to include LEDs. For instance, if you want to turn your normal jacket into an advertisement for HowStuffWorks, you'd send the jacket and a description of what you wanted to the alteration company. A company employee would cut small holes in the jacket and fit an LED into each hole.

Once assembled, the lights spell out your message. On the inside of the jacket, each LED attaches to a circuit board. Most companies cover the circuitry with a thin fabric lining. The employee then attaches a power source, usually a small battery pack that can fit into a jacket, shirt, or pants pocket.

Other companies cover the LEDs with a thin fabric so that the plastic bulbs aren't visible. When turned on, the light shines through the thin cloth. Like EL clothing, clothes using LED technology must be carefully maintained and washed. Otherwise, the delicate circuitry could become damaged.

The circuitry controlling an LED display can be as simple as a power switch, meaning all the LEDs are either turned on or off, or it can include microprocessors that let the wearer customize how the LEDs switch on and off. This means that if a shirt has a block of RGB LED lights (LEDs that can emit any color of light), the wearer can create a program telling each LED when to turn on and off as well as what color it should be. By alternating colors and turning on and off, an LED fabric display can create simple images or messages. Each LED acts as a pixel, but because LEDs are much larger than television pixels, the resolution on an LED display won't be very sharp. Because of the low resolution, clothes with LEDs are best for simple messages or designs.

Using LED displays, you can have a preprogrammed light show on your clothing. But what if you want an even more impressive display? In that case, you'll need to look at technology that can display full moving pictures like television signals.

Let's look closely at one of these technologies—PLEDs.

⫷ FURRY TVS? FABRIC DISPLAYS USING PLEDS

Polymer light emitting diode (PLED) is a technology used in backlighting, illumination, and electronic displays. Unlike LEDs, which are small bulbs, a PLED display is a thin, flexible film made of polymers and capable of emitting the full color spectrum of light.

A PLED is constructed of several layers:

- An engineer begins with a glass or plastic substrate—for PLED fabric displays, plastic tends to be a better choice because it's less fragile and more flexible than glass.
- Next comes a transparent electrode coating, which an engineer applies to one side of the substrate.
- Then the engineer coats the same side of the substrate with the light-emitting polymer film.
- The last layer is an evaporated metal electrode, which the engineer applies to the other side of the polymer film.

When the engineer applies an electric field between the two electrodes, the polymer emits light, much like an LED. Because the polymers in PLED are made of organic molecules, they are also known as organic light emitting diodes (OLEDs).

Using a PLED screen, it would be possible to create a fabric television. PLED displays are very thin and relatively light compared to other display technologies. Of course, the screen is just one important element in the overall fabric display—you would also need a power source, such as a lithium-ion battery, and a signal source. The signal source could be a small computer containing preloaded video clips or even a Wi-Fi-enabled device that could stream audio and video directly to your clothes.

Clothes using PLED displays aren't currently available, though several web pages list fabric displays as a likely PLED application in the near future.

Next we'll read about how an invention called T-Shirt TV works and the impact it's made on advertising.

WHAT'S A POLYMER?

Polymers are chains of repeated molecular structures. The shape of the molecular structure determines the polymer's properties. For example, graphite and diamonds are both inorganic carbon polymers—they have different properties because of the way the carbon atoms bind together to form molecules. The molecular structure of the polymers in PLEDs is what gives them the ability to emit light after electrical stimulation.

▌▌▌▐ T-SHIRT TELEVISION

In 2004, Brand Marketers introduced Adver-Wear, shirts with built-in 11-inch (28-centimeter) television screens and a four-speaker sound system. Adam Hollander designed and produced the first Adver-Wear shirts—also known as T-Shirt TV—because he felt that static advertisements didn't reach younger consumers. In interviews, Hollander said that young people had become so used to television that marketers needed to find new ways to incorporate video and animation in advertising strategies or risk losing customers.

The shirts weigh about six and a half pounds (nearly three kilograms). Because they use flat-panel television screens, the shirts are a little bulky. Hollander created the shirts specifically for advertising, with no intention of offering them to the public, so comfort and practicality weren't of much concern in the design process. The components are also too expensive to sell T-Shirt TV clothing to the consumer market.

In addition to the screen and speaker system, the shirt needs a portable power source—usually in the form of a lithium-ion battery. It also needs a video source—a custom-built computer that can store and play digital media efficiently. Brand Marketers can incorporate the system into most types of clothes as well as include interactive features like a handheld touch screen if clients request it. Currently, the system can play any length of audio or video files separately or in a loop. Future versions of the shirt will include Bluetooth and Wi-Fi capability, adding more interactive features to the clothing.

The marketing firm Adwalkers offers a similar product: a portable, wearable computer system. A padded harness acts as the frame for the system, which includes a portable PC, computer monitor, handheld touch screen, and printer. Clients can send Adwalkers videos, software, logos, and other computer files, and Adwalkers incorporates them into eye-catching presentations. An Adwalkers employee wears the harness and interacts with the general public. Using this system, the Adwalkers employee can:

- Display graphics like company logos or screenshots of web pages on the computer monitor.
- Play video and audio.
- Run interactive computer software, including games, which members of the public can access using the handheld touch screen.
- Print tickets, coupons, or brochures for customers.
- Capture customer data using spreadsheet or database software.
- Interact with the public and answer questions about the advertised product.

The system is bulkier than Adver-Wear, but it's even more interactive and has more functions than T-Shirt TV.

As advertising companies continue to search for new ways to grab our attention, we'll likely see more applications of fabric displays. We may even see applications where entire outfits act like a television screen, with images wrapping around from front to back.

COMING TO A T-SHIRT NEAR YOU

Adver-Wear made a big splash during the promotional campaign for the Will Smith film *I, Robot*. Models in several different cities wore Adver-Wear shirts with screens playing a looped preview for the film. The marketing campaign caught national news attention.

BACK TO THE FUTURE: FIVE FUTURE TECHNOLOGY MYTHS

From what we've learned so far, the future certainly has a lot of awesome things in store for us! But we've heard about a lot of fairly outlandish stuff, too. What are we supposed to believe?

We began this book with futurists, so it seems only appropriate to circle back to these harbingers of things to come. To wrap up this book, let's take a final look at what the world will look like ten years from now, forty years from then, and beyond. Will we live in a world run by robots? Will we all be driving flying cars? Will we have conquered global warming? We'll explore some popular ideas about the future of technology that are likely myths.

Predicting future trends or developments, especially in a dynamic field like technology, is inherently inexact, but it is possible to make some informed guesses. We looked at some developments earlier because there is an argument for their plausibility, but in other cases, there's enough evidence out there, particularly from experts, to diagnose them as myths.

Let's start with one of the great fabled machines of the post-industrial age: the flying car.

1. SOON WE'LL ALL BE DRIVING FLYING CARS

The flying car has been prophesied for decades. It's the holy grail of the futuristic, utopian society, where everyone gets to zip around through the air and land easily, quietly, and safely wherever he or she wants.

You've probably seen videos of flying-car prototypes taking off from the ground, hovering, and possibly crashing. But the first "autoplane" was actually unveiled in 1917, and many similar efforts have followed. Henry Ford predicted the flying car was coming—in 1940—and there have been numerous false alarms ever since.

A decade into the twenty-first century, we don't seem to be any closer, despite what you might read on gadget blogs. Because funding dried up, NASA abandoned its contest for inventors to create a "personal air vehicle," and there doesn't seem to be another government agency, except perhaps the secretive DARPA, ready to take on the project.

There are simply too many challenges in the way of a flying car becoming widely adopted. Cost, flight paths and regulations, safety, potential use in terrorism, fuel efficiency, training pilots (drivers), landing, noise, and opposition from the automobile and transportation industries all stand in the way of a legitimate flying car. Also, these vehicles will likely have to be able to operate as cars on regular roads, posing another logistical challenge.

˙In fact, many of the so-called flying cars that are being hawked as the real thing are simply roadable aircrafts—a sort of plane-car hybrid that is not even capable of, for example, making a short trip to school to drop off the kids. Plus, they're far too expensive. One such vehicle, the Terrafugia Transition, completed its first test flight on October 27, 2012. The expected base price is $279,000.

2. THE TECHNOLOGICAL SINGULARITY APPROACHES

In recent years, prominent futurists like Ray Kurzweil have argued that we are approaching "the singularity," perhaps as soon as 2030. There are many different conceptions of just what exactly the singularity is or will be. Some say it's a true artificial intelligence that can rival humans in independent thinking and creativity. In other words, machines will surpass humans in intelligence and, as the planet's dominant species, be capable of creating their own new, smarter machines. Others contend that it will involve such an explosion in computing power that somehow humans and machines will merge to create something new, such as by uploading our minds onto a shared neural network.

Critics of the singularity, such as writer and academic

Douglas Hofstadter, claim that these are "science-fiction scenarios" that are essentially speculative. Hofstadter calls them vague and useless in contemporary discussions of what makes a human being and our relationships with technology. There is also little evidence that the sort of "tidal wave" of technological innovation predicted by Kurzweil and other futurists is imminent.

Mitch Kapor, the former CEO of Lotus, called the singularity "intelligent design for the IQ 140 people." One magazine called it "the Rapture for the Geeks"—hardly a complimentary term. Computer scientist Jeff Hawkins contends that while we may create highly intelligent machines—far greater than anything we have now—true intelligence relies on "experience and training," rather than just advanced programming and processing power.

Doubters point to the numerous sci-fi fantasies and predictions of the past that still have not come true as evidence that the singularity is just another pie-in-the-sky dream—for example, we don't have moon bases or artificial gravity yet. They also argue that understanding the nature of consciousness is impossible, much less creating this capability within machines. Finally, the impending coming of the singularity depends in large part on the continuation of Moore's law, which, as we discuss next, may be in jeopardy. (It should also be noted that Gordon Moore is not a believer in the singularity.)

3. MOORE'S LAW WILL ALWAYS HOLD TRUE

We hinted earlier that this may not be the case, but let's take a deeper look. If you recall, Moore's law is generally taken to mean that the number of transistors on a chip—and by extension, processing power—doubles every two years. In reality, Gordon Moore, the computer scientist who originated Moore's law in 1965, was talking about the economic costs of chip production and not the scientific achievements behind advances in chip design.

Moore believed that the costs of chip production would halve annually for the next ten years but may not be sustainable afterward. The limit to Moore's law may then be reached economically instead of scientifically.

Several prominent computer experts have contended that Moore's law cannot last more than two decades. Why is Moore's law doomed? Because chips have actually become much more expensive to produce as transistors have become smaller.

The factories that produce these chips cost billions of dollars. GLOBAL FOUNDRIES's Fab 8 factory, which began production in New York in 2012, cost $4.2 billion to build. Few companies have those kinds of resources, and Intel has said that a company must have $9 billion in yearly revenue to compete in the cutting-edge chip market.

Companies may also attempt to make the most out of

current technologies before investing in new, more expensive, smaller chip designs. So while the end of Moore's law may limit the rate at which we add transistors to chips, that does not necessarily mean that other innovations will prevent the creation of faster, more advanced computers.

IIII| 4. ROBOTS WILL BE OUR FRIENDS

We're probably not headed for a Skynet-like Armageddon, nor are we doomed to a future resembling *The Matrix*, but an increasing number of scientists worry whether adequate measures are being taken to safeguard ourselves from our robotic and digital creations.

One of the main concerns is automation. Will military drones eventually be allowed to make their own decisions on whether or not to attack a target? If a human is monitoring, will he or she still be able to override the drone's wishes? Will we allow machines to replicate themselves without human direction? Are we going to allow self-driving cars? (Some cars already offer the ability to park themselves or to prevent a driver from drifting into another lane.)

Then there is the issue of robots occupying roles they probably should not. Already, there are prototype medical robots designed to ask patients about their symptoms and to provide counsel, simulating comforting emotions—a role traditionally performed by a human doctor. Microsoft has a video-based receptionist, A.I., in one of its buildings. A new class of "service robots" can plug themselves into electrical outlets and perform other menial tasks—not to mention the long-established Roomba, an automated, vacuuming robot.

We may also be placing too many critical tasks and responsibilities into the "hands" of nonhuman actors, or will gradually find ourselves in a position of dependence on machines. At a 2009 conference of computer scientists, roboticists, and other researchers, the experts in attendance expressed concern about how criminals could take advantage of next-generation technology, like artificial intelligence, to hack information or impersonate real people.

The bottom line of this conference and other discussions seems to be that it's important to start tackling these issues early, to outline industry standards now, even if it's not clear what kind of technological advancements the future will bring.

THE END OF WORK?

HowStuffWorks founder Marshall Brain has written about his fear that the widespread adoption of robots for jobs traditionally occupied by humans could lead to mass unemployment.

▥ 5. WE CAN STOP CLIMATE CHANGE

Is global warming inevitable? The consensus among many scientists is that it is, at least to some extent, and that we can only hope to stop major disasters and deal with the consequences. Some of the world's most respected climatologists say that humanity has already passed the proverbial point of no return. The UN Intergovernmental Panel on Climate Change, a group of more than two thousand scientists, met in 2007 and issued a stark warning, after having first announced in 2001 that global temperatures were already rising.

Even now, we are seeing the effects of climate change, such as glacier melt and rising sea levels making South Asian cyclones more severe. The effects are expected to be particularly severe for hundreds of millions of people in the developing world. The atoll of Tuvalu now deals with high tides that threaten to submerge the entire nation.

If we produced no more greenhouse gases after today, the world would still see a 1°F (0.6°C) increase in temperature by midcentury because existing carbon dioxide would stay in the atmosphere for a half-century or more. (Some countries are trying to do something about this, such as Norway, which is pumping CO_2 into disused underground oil wells.) And a potentially catastrophic increase of 3 to 6°F (1.7 to 3.3°C) by the end of the century is possible.

The major remaining question, for some, is whether the amount of warming can be kept in check to prevent these disastrous scenarios. Encouraging grassroots environmental action is important, but intergovernmental cooperation is paramount, and that's been slow in coming, particularly with the United States, China, and India. Experts say we also need to begin to plan how to respond to warming-related disasters by aiding coastal areas, establishing quick-response units for wildfires, and preparing for deadly heat waves.

What do you think, reader? Will our future selves scoff at these myths, or will they be borne out? If only we could climb into a time-traveling machine and find out.

CONCLUSION

I f we take a step back and look at everything we've learned about the future, we can expect some pretty cool stuff in the decades to come! From digital immortality and robotic assistants to homes that can practically clean themselves and virtual reality games we can literally step into, the future features a lot of exciting possibilities. Of course, there are upsides and downsides to every advancement, and as always, we have to be evermore vigilant about safeguarding ourselves and our identities. We also need to be far more proactive about protecting our planet. If we can achieve both of these goals and keep developing new technology like some of the ideas we've explored here, we can look forward to a fantastic future.

SOURCES

IIII¡I PART I: MODERN-DAY FORTUNE-TELLERS: HOW FUTUROLOGY AND FUTURISTS WORK

Acceleration Watch. "Futurist." Accessed August 7, 2012. accelerationwatch.com/futuristdef.html.

Association of Professional Futurists. "What Is a Futurist?" Accessed August 7, 2012. www.profuturists.org/futurists.

Chalupa, Andrea. "Hearing the Buzz Before It's Buzz." *Upstart Business Journal*, September 4, 2007. Accessed August 7, 2012. upstart .bizjournals.com/careers/job-of-the-week/2007/09/04/The -Futurist.html?page=all.

Cornish, Edward. "The Futurist Interviews Sir Arthur C. Clarke." World Future Society. Accessed August 7, 2012. www.wfs.org/node/852.

Crawford, Mathias. "What Futurists Actually Do." *Good*, July 13, 2010. Accessed August 7, 2012. www.good.is/post/what-futurists-actually-do.

Eaves, Elisabeth. "The Futurists." *Forbes*, October 15, 2007. Accessed August 8, 2012. www.forbes.com/2007/10/13/futurist -business-consultant-tech-future07-cx_ee_1015futurist.html.

Future Search. "Futurist FAQ." Accessed August 7, 2012. www .futuresearch.com/faq.php.

"Futurist." *Oxford English Dictionary*. Accessed August 23, 2012. www .oed.com.

Glasner, Joanna. "The Future Needs Futurists." *Wired*, October 7, 2005.

Accessed August 7, 2012. archive.wired.com/techbiz/media /news/2005/10/69087.

Global Future Report. "What Is a Futurist, Exactly?" March 2, 2005.

Institute for Global Futures. "James Canton, PhD: Futurist, Author, and Visionary Business Advisor." Accessed August 7, 2012. www.globalfuturist.com/dr-james-canton/biography .html?gclid=CMqD3tes1rECFce5KgodAiwAEQ.

Marshall, Michael. "Five Futurist Visionaries and What They Got Right." *New Scientist*, May 6, 2009. Accessed August 7, 2012. www .newscientist.com/article/dn17082-five-futurist-visionaries -and-what-they-got-right.html.

Mullins, John. "Futurist." U.S. Department of Labor, Bureau of Labor Statistics. Spring 2009. Accessed August 7, 2012. www.bls.gov /opub/ooq/2009/spring/yawhat.htm.

O'Leary, Siobhan. "Arthur C. Clarke: A Futurist Who Got It Right (and Predicted the iPad)." *Publishing Perspectives*, September 6, 2010. Accessed August 7, 2012. publishingperspectives .com/2010/09/arthur-c-clarke-a-futurist-who-got-it-right -and-predicted-the-ipad.

Ryan, Ellen. "What's Ahead: A Futurist Predicts." *Washingtonian*, February 1, 2008. Accessed August 7, 2012. www.washingtonian.com /articles/people/whats-ahead-a-futurist-predicts.

| | | | |

FIVE FUTURIST PREDICTIONS IN THE WORLD OF TECHNOLOGY

Ackerman, Evan. "DARPA Wants to Give Soldiers Robot Surrogates, Avatar Style." *IEEE Spectrum*, February 17, 2012. Accessed August 13, 2012. spectrum.ieee.org/automaton/robotics/military-robots /darpa-wants-to-give-soldiers-robot-surrogates-avatar-style.

Bergin, Chris. "NASA Exploration Roadmap: A Return to the Moon's
    Surface Documented." NASASpaceFlight.com. March
    19, 2012. Accessed August 12, 2012. www.nasaspaceflight
    .com/2012/03/nasa-exploration-roadmap-return-moons
    -surface-documented.

Carmichael, Mary. "Neuromarketing: Is It Coming to a Lab Near You?"
    PBS. November 9, 2004. Accessed August 11, 2012. www.pbs
    .org/wgbh/pages/frontline/shows/persuaders/etc/neuro.html.

Dillow, Clay. "Microsoft's 'Universal Translator' Lets You Speak
    Foreign Languages in Your Own Voice." *Popular Science*,
    March 12, 2012. Accessed August 12, 2012. www.popsci.com
    /technology/article/2012-03/microsofts-universal-translator
    -converts-speech-while-preserving-accent-and-timbre.

Fantz, Ashley. "Who Is Anonymous? Everyone and No One."
    CNN. February 9, 2012. Accessed August 11, 2012.
    articles.cnn.com/2012-02-09/world/world_anonymous
    -explainer_1_chat-room-internet-caf-anonymous-members
    ?_s=PM:WORLD.

Institute for Global Futures. "Global Futures Forecast 2012." Accessed August
    10, 2012. www.globalfuturist.com/images/docs/GFF2012.pdf.

Landau, Elizabeth. "What We've Done on Mars and What's Next."
    CNN. August 12, 2012. Accessed August 13, 2012. www.cnn
    .com/2012/08/11/tech/innovation/mars-exploration-history
    /index.html.

Massachusetts Institute of Technology. "How to Build a Phononic
    Computer." CNN. August 13, 2012. Accessed August 13, 2012.
    www.technologyreview.com/view/428844/how-to-build
    -a-phononic-computer.

Menegaz, Gery. "Moore's Law: The End Is Near-ish!" ZDNet. July 16,
    2012. Accessed August 10, 2012. www.zdnet.com/moores
    -law-the-end-is-near-ish-7000000972.

National Aeronautics and Space Administration (NASA). "Space Launch System." Accessed August 13, 2012. www.nasa.gov /pdf/664158main_sls_fs_master.pdf.

National Nanotechnology Initiative. "What Is Nanotech?" Accessed August 10, 2012. www.nano.gov.

Sarchet, Penny. "Nanofactories: A Future Vision." *The Guardian*, November 25, 2011. Accessed August 11, 2012. www.guardian .co.uk/nanotechnology-world/nanofactories-a-future-vision.

Seligson, Joelle. "Q&A with Brian David Johnson." Association of Science-Technology Centers. Accessed August 10, 2012. www.astc.org/pubs/dimensions/2012/Jul-Aug/Q&A_Brian DavidJohnsonTranscript.pdf.

The Space Elevator Reference. Accessed August 10, 2012. www .spaceelevator.com.

| | | | |

‖‖ COGNITIVE CRAZINESS: HOW COGNITIVE TECHNOLOGY WILL CHANGE OUR LIVES

Dascal, Marcelo. "Language as a Cognitive Technology." Tel Aviv University. Accessed August 7, 2012. www.tau.ac.il /humanities/philos/dascal/papers/ijct-rv.htm.

Dillow, Clay. "Will People Alive Today Have the Opportunity to Upload Their Consciousness to a New Robotic Body?" *Popular Science*, March 2, 2012. Accessed August 7, 2012. www.popsci.com/technology/article/2012-03/achieving -immortality-russian-mogul-wants-begin-putting-human -brains-robots-and-soon.

Dror, Itiel and Stevan R. Harnad. *Cognition Distributed: How Cognitive Technology Extends Our Minds*. Philadelphia: John Benjamins Publishing Co., 2008.

Feiner, Steven, et al. "MARS: Mobile Augmented Reality Systems."
Columbia University Computer Graphics and User Interfaces
Lab. Accessed August 7, 2012. graphics.cs.columbia.edu
/projects/mars.

Walker, W. Richard and Douglas J. Herrmann. *Cognitive Technology:
Essays on the Transformation of Thought and Society*. Jefferson, NC:
McFarland and Co., 2005.

| | | | |

EDUCATION: HOW WILL FUTURE TECHNOLOGY CHANGE THE CLASSROOM?

Dede, Chris and John Richards. "Customizing the Classroom for
Each Student Using Digital Resources." *District Administration*,
June 2012.

Grantham, Nick. "Five Future Technologies That Will Shape Our
Classrooms." Edutopia. April 10, 2012. Accessed August 24, 2012.
www.edutopia.org/blog/five-future-education-technologies
-nick-grantham.

Hopkins, Curt. "Future U: Classroom Tech Doesn't Mean Handing Out
Tablets." Ars Technica. May 6, 2012. Accessed August 24, 2012.
arstechnica.com/features/2012/05/future-u-classroom-tech.

Human-Computer Interaction Lab. "Classroom of the Future." University
of Maryland. Accessed August 24, 2012. www.cs.umd.edu
/hcil/kiddesign/cof.shtml.

Ledesma, Patrick. "Can You Predict the Future Technologies in Your
Classroom?" *Education Week*, February 21, 2011. Accessed
August 24, 2012. blogs.edweek.org/teachers/leading_from
_the_classroom/2011/02/can_you_predicting_the_future
_technologies_in_your_classroom.html.

National Center for Education Statistics. "Fast Facts." Institute of

Education Sciences. Accessed August 24, 2012. nces.ed.gov /fastfacts/display.asp?id=46.

Pandolfo, Nick. "Education Technology: As Some Schools Plunge In, Poor Schools Are Left Behind." *The Hechinger Report*, January 24, 2012. Accessed August 24, 2012. www.huffingtonpost .com/2012/01/24/education-technology-as-s_n_1228072 .html.

Reilly, Michael. "The Intelligent Textbook That Helps Students Learn." *NewScientist*, August 7, 2012. Accessed August 24, 2012. www .newscientist.com/article/mg21528765.700-the-intelligent -textbook-that-helps-students-learn.html.

Richtel, Matt. "In Classroom of Future, Stagnant Scores." *New York Times*, September 3, 2011. Accessed August 24, 2012. www .nytimes.com/2011/09/04/technology/technology-in -schools-faces-questions-on-value.html?pagewanted=all.

Time to Know. Accessed August 24, 2012. www.timetoknow.com.

| | | | |

FUN IN THE FUTURE: WHAT ENTERTAINMENT WILL LOOK LIKE IN 2050

Berry, James R. "What Will Life Be Like in the Year 2008?" *Mechanix Illustrated*, November 1968. Accessed April 9, 2009. blog .modernmechanix.com/2008/03/24/what-will-life-be-like -in-the-year-2008.

Choi, Charles Q. "Forecast: Sex and Marriage with Robots by 2050." LiveScience. October 12, 2007. Accessed April 9, 2009. www .livescience.com/technology/071012-robot-marriage.html.

Della Cava, Marco R. "Healthy, Aged Boomers Could Dominate 2046 Landscape." *USA Today*, October 27, 2005.

Edwards, John. "The Internet of Things." *RFID Journal*, April 23,

2012. Accessed August 7, 2012. www.rfidjournal.com/article
/purchase/9424.

Gutkind, Lee. "Bend It Like Robo-Beckham." Salon. June 11,
2003. Accessed April 9, 2009, dir.salon.com/story/tech
/feature/2003/06/11/robocup/index.html.

Johnson, Alex. "America in 2050: Even Older and More Diverse."
MSNBC. August 14, 2008. Accessed April 9, 2009. www
.msnbc.msn.com/id/26186087.

Naisbitt, John. "The Postliterate Future." *Futurist*, March–April 2007.

Newsweek. "How Tweet It Is." June 9, 1997.

New York Times. "A Forecast for 2050: Scarcities Will Force a Leaner
U.S. Diet." February 18, 1995. Accessed April 9, 2009. www
.nytimes.com/1995/02/18/us/a-forecast-for-2050-scarcities
-will-force-a-leaner-us-diet.html?scp=9&sq=2050&st=cse.

Pacotti, Sheldon. "Are We Doomed Yet?" Salon. March 31,
2003. Accessed April 9, 2009. dir.salon.com/story/tech
/feature/2003/03/31/knowledge/index.html.

Pearce, Fred. "No More Seafood by 2050?" *NewScientist*, November 2,
2006. Accessed April 9, 2009. www.newscientist.com/article
/dn10433-no-more-seafood-by-2050.html.

Peterson, Dan. "Group Predicts Robots Will Win Soccer's World Cup in
2050." LiveScience. February 26, 2009. Accessed April 9, 2009.
www.livescience.com/technology/090226-robocup-robots
-soccer.html.

Sakr, Sharif. "DARPA Realizes It Needs Contact Lenses, Opts for Those
Nice AR Tinted Ones." Engadget. April 13, 2012. Accessed
August 7, 2012. www.engadget.com/2012/04/13/darpa
-innovega-ioptik-augmented-reality-contact-lenses.

Smith, David. "2050 and Immortality Is Within Our Grasp." *The Observer*,
May 21, 2005. Accessed April 9, 2009. www.guardian.co.uk
/science/2005/may/22/theobserver.technology.

Sutherland, Ivan E. "The Ultimate Display." U.S. Department of Defense, Advanced Research Projects Agency, 1965. Accessed August 7, 2012. www.eng.utah.edu/~cs6360/Readings/UltimateDisplay.pdf.

Williams, Caroline. "Jellyfish Sushi: Seafood's Slimy Future." *NewScientist*, March 4, 2009. Accessed April 9, 2009. www .newscientist.com/article/mg20126981.900-jellyfish-sushi -seafoods-slimy-future.html.

Wilson, Jim. "Miracles of the Next 50 Years." *Popular Mechanics*, February 2000.

| | | | |

IIIIıI HAPPY AND HEALTHY? HOW MEDICINE AND OUR HEALTH WILL CHANGE

Association of Professional Futurists. "What Is a Futurist?" Accessed August 10, 2012. www.profuturists.org/futurists.

Atala, Anthony. "Regenerative Medicine's Promising Future." CNN. July 10, 2011. Accessed August 10, 2012. articles.cnn .com/2011-07-10/opinion/atala.grow.kidney_1_regenerative -organ-transplants-new-organs?_s=PM:OPINION.

Buntz, Brian. "Theoretical Physicist Michio Kaku Predicts the Future of Healthcare." *Medical Device and Diagnostic Industry*, November 10, 2011. Accessed August 9, 2012. www .mddionline.com/article/theoretical-physicist-michio-kaku -predicts-future-healthcare.

Carroll, Jim. "Trend Report: The Future of Health, Fitness, and Wellness." July 17, 2012. Accessed August 10, 2012. www .jimcarroll.com/category/trends/health-care-trends.

Centers for Disease Control and Prevention. "Heart Disease Facts." March 23, 2012. Accessed August 9, 2012. www.cdc.gov /HeartDisease/facts.htm.

Devlin, Kate. "British Doctors Help Perform World's First Transplant of a Whole Organ Grown in a Lab." *The Telegraph*, November 18, 2008. Accessed August 10, 2012. www.telegraph.co.uk /health/healthnews/3479613/British-doctors-help-perform -worlds-first-transplant-of-a-whole-organ-grown-in-lab.html.

Flower, Joe. "Better Ways of Thinking about the Future." Imagine What If. January 24, 2012. Accessed August 10, 2012. www .imaginewhatif.com/better-ways-of-thinking-about-the-future.

Flower, Joe. "On Beyond Healthcare: Save the Country with Preventive Care." Imagine What If. May 23, 2012. Accessed August 9, 2012. www.imaginewhatif.com/on-beyond-healthcare-save -the-country-with-preventive-care.

Gold, Jenny. "Accountable Care Organizations, Explained." NPR. January 18, 2011. Accessed August 10, 2012. www.npr.org/2011 /04/01/132937232/accountable-care-organizations-explained.

Guttmacher, Alan E. and Francis S. Collins. "Genomic Medicine: A Primer." *New England Journal of Medicine* 347 (November 7, 2002): 1512–1520. Accessed August 10, 2012. www.nejm.org /doi/full/10.1056/NEJMra012240.

Khan, Ilyas. "The Future Is Behind Us: Tissue Engineering, the State of the Art." Cardiff University, School of Biosciences. Accessed August 10, 2012. cardiffsciscreen.blogspot.com/2011/03 /future-is-behind-us-tissue-engineering.html.

LeVaux, Ari. "Experimenting with Nootropics to Increase Mental Capacity, Clarity." *The Atlantic*, January 30, 2012. Accessed August 8, 2012. www.theatlantic.com/health /archive/2012/01/experimenting-with-nootropics-to -increase-mental-capacity-clarity/252162/#.

Maher, Brendan. "Poll Results: Look Who's Doping." *Nature*, April 9, 2008. Accessed August 8, 2012. www.nature.com /news/2008/080409/full/452674a.html.

McClanahan, Carolyn. "How Much Should We Spend on Health Care? The Big Picture." *Forbes*, November 28, 2011. Accessed August 10, 2012. www.forbes.com/sites /carolynmcclanahan/2011/11/28/how-much-should-we -spend-on-health-care-the-big-picture.

McGowan Institute for Regenerative Medicine. "What Is Regenerative Medicine?" 2010. Accessed August 10, 2012. regenerativemedicine.net/What.html.

Miller, Laura. "Healthcare Futurist Joe Flower: 5 Best Practices to Prepare Hospitals for Accountable Care." *Beckers Hospital Review*, September 21, 2010. Accessed August 12, 2012. www .beckershospitalreview.com/hospital-physician-relationships /healthcare-futurist-joe-flower-5-best-practices-to-prepare -hospitals-for-accountable-care.html.

Moss, Frank. "Our High-Tech Health Care Future." *New York Times*, November 9, 2011. Accessed August 9, 2012. www.nytimes .com/2011/11/10/opinion/our-high-tech-health-care-future .html?_r=1.

PBS. "The Bionic Body: The Body Shop." *Scientific American Frontiers*, 2011. Accessed August 9, 2012. www.pbs.org/saf/1107 /features/body.htm.

Pelletier, Dick. "Regenerative Medicine Could Cure Most Diseases by 2020." Positive Futurist. Accessed August 9, 2012. www .positivefuturist.com/archive/111.html.

ScienceDaily. "New Concept for Fast, Low-Cost DNA Sequencing Device." Oak Ridge National Laboratory, April 24, 2012. Accessed August 11, 2012. www.sciencedaily.com /releases/2012/04/120424120455.htm.

Tandon, Aja, et al. "Measuring Overall Healthcare Performance for 191 Countries." World Health Organization. Accessed August 10, 2012. www.who.int/healthinfo/paper30.pdf.

Uldrich, Jack. "Top Ten Healthcare Trends." Jump the Curve. October 13, 2011. Accessed August 9, 2012. jumpthecurve.net/health-care/top-ten-healthcare-trends-by-futurist-jack-uldrich.

Winslow, Ron and Shirley S. Wang. "Soon $1,000 Will Map Your Genes." *Wall Street Journal*, January 10, 2012. Accessed August 10, 2012. online.wsj.com/article/SB10001424052970204124204577151053537379354.html.

| | | | |

IIIIIII DIGITAL DOCTORS: WILL COMPUTERS REPLACE DOCTORS?

Bottles, Kent. "Will Patients Trust Sociable Human Robots?" KevinMD.com. Accessed August 15, 2012. www.kevinmd.com/blog/2011/08/patients-trust-sociable-humanoid-robots.html.

Brubaker, Michelle. "Text4baby Mobile Service Shows Positive Result for New Moms." University of California, San Diego, Communications and Public Affairs. November 14, 2011. Accessed August 1, 2012. ucsdnews.ucsd.edu/pressreleases/text4baby_mobile_service_shows_positive_results_for_new_moms/#.UBsvOLRDySo.

ERtexting. Accessed August 1, 2012. www.ertexting.com.

Fachot, Morand. "Doctor Robot, I Presume? International Electrotechnical Commission. July 2011. Accessed August 15, 2012. www.iec.ch/etech/2011/etech_0711/ind-2.htm.

IBM. "Putting Watson to Work." Accessed August 15, 2012. www-03.ibm.com/innovation/us/watson/watson_in_healthcare.shtml.

Klein, Ezra. "How Robots Will Replace Doctors." *Washington Post*, October 1, 2011. Accessed August 15, 2012. www.washingtonpost.com/blogs/ezra-klein/post/how-robots-will-replace-doctors/2011/08/25/gIQASA17AL_blog.html.

Manjoo, Farhad. "Will Robots Steal Your Job?" Slate. September 27,

2011. Accessed August 15, 2012. www.slate.com/articles /technology/robot_invasion/2011/09/will_robots_steal _your_job_3.single.html.

Marquez, Jose. "Will mHealth Revolutionize Healthcare?" Huffington Post. March 7, 2012. Accessed August 2, 2012. www .huffingtonpost.com/jose-marquez/will-mhealth-revolutioniz _b_1324991.html.

Medic Mobile. Accessed August 1, 2012. medicmobile.org.

Sapkota, Nabin. "Computers Won't Replace Doctors." *Columbus Telegram*, May 22, 2012. Accessed August 15, 2012. columbustelegram .com/news/local/computers-won-t-replace-doctors/article _b0fdd4de-a411-11e1-942c-0019bb2963f4.html.

Sweeney, Chris. "How Text Messages Could Change Global Healthcare." *Popular Mechanics*, October 24, 2011. Accessed August 2, 2012. www.popularmechanics.com/science/health/med-tech/how -text-messages-could-change-global-healthcare.

Tanner, Lindsey. "Doctors Using Social Media; Tweeting and Texting Patients." Globalnews.ca. June 11, 2012. Accessed August 1, 2012. globalnews.ca/news/254809/doctors-using-social -media-tweeting-and-texting-with-patients.

U.S. Department of Health and Human Services. "Health Text Messaging Recommendations to the Secretary." 2010. Accessed August 1, 2012. www.hhs.gov/open/initiatives/mhealth/recommendations .html.

U.S. National Library of Medicine. "Robotic Surgery." Medline Plus. 2011. Accessed August 15, 2012. www.nlm.nih.gov /medlineplus/ency/article/007339.htm.

IIII‖I MIND OVER MATTER: TELEKINESIS, MIND CONTROL, AND DIGITAL IMMORTALITY

American Pain and Wellness. "Spinal Injections Procedures: Neurostimulation." Accessed August 8, 2012.

Aron, Jacob. "Kinect Lab Boss on the Future of Computer Interfaces." *NewScientist*, March 28, 2012. Accessed August 8, 2012. www .newscientist.com/article/dn21634-kinect-boss-on-the-future-of -computer-interfaces.html?DCMP=OTC-rss&nsref=online-news.

Boyle, Rebecca. "New Computer Chip Modeled on a Living Brain Can Learn and Remember." *Popular Science*, August 18, 2011. Accessed August 8, 2012. www.popsci.com/technology /article/2011-08/first-generation-cognitive-chips-based-brain -architecture-will-revolutionize-computing-ibm-says.

Chudler, Eric H. "Brain Facts and Figures." University of Washington, Neuroscience for Kids. Accessed August 8, 2012. faculty .washington.edu/chudler/facts.html.

Chudler, Eric H. "Neurotransmitters and Neuroactive Peptides." University of Washington, Neuroscience for Kids. Accessed August 21, 2009. faculty.washington.edu/chudler/chnt1.html.

Dillow, Clay. "Will People Alive Today Have the Opportunity to Upload Their Consciousness to a New Robotic Body?" *Popular Science*, March 2, 2012. Accessed August 8, 2012. www.popsci.com /technology/article/2012-03/achieving-immortality-russian -mogul-wants-begin-putting-human-brains-robots-and-soon.

The Economist. "Neuroscience: Sound and No Fury," January 8, 2009. Accessed August 21, 2009. www.economist.com/science technology/displayStory.cfm?story_id=12887217&fsrc=rss.

Jauchem, James R. "High-Intensity Acoustics for Military Nonlethal Applications: A Lack of Useful Systems." *Military Medicine* 172 (February 2007): 182–189. Accessed March 7, 2014. www .ncbi.nlm.nih.gov/pubmed/17357774.

Nature. "Brain Machine Interfaces." Accessed August 7, 2012. www
.nature.com/nature/focus/brain.

Nature. "Editorial: Is This the Bionic Man?" July 13, 2006. Accessed
August 8, 2012. www.nature.com/nature/journal/v442
/n7099/full/442109a.html.

ScienceDaily. "Does the Brain Control Muscles or Movements?"
Cell Press. May 8, 2008. Accessed August 8, 2012. www
.sciencedaily.com/releases/2008/05/080507133321.htm.

ScienceDaily. "Ultrasound Shown to Exert Remote Control of
Brain Circuits." Arizona State University. November 2,
2008. Accessed August 21, 2009. www.sciencedaily.com
/releases/2008/10/081029104251.htm.

Song, Sora. "How Deep-Brain Stimulation Works." *Time*, July 16, 2006.
Accessed August 21, 2009. www.time.com/time/magazine
/article/0,9171,1214939,00.html.

Ter Haar, Gail and Constantin Coussios. "High-Intensity Focused
Ultrasound: Past, Present, and Future." *International Journal of
Hyperthermia* 23 (March 2007): 85–87.

Williams, Robert W. and Karl Herrup. "The Control of Neuron
Number." *Annual Review of Neuroscience* 11 (September 28,
2001): 423–453. Accessed August 25, 2009. www.nervene
t.org/papers/NUMBER_REV_1988.html#1.

| | | | |

IIII| COMPUTING OUR CULTURE: SUPERCOMPUTING TO MAKE US SUPERHUMANS

Daily Mail. "Get Ready for the Supercomputer That Can Predict the
Future." December 5, 2011. Accessed August 8, 2012. www
.dailymail.co.uk/sciencetech/article-2069775/Get-ready-super
computer-predict-future-EU-prepares-900m-funding.html.

Graham, Susan L., Marc Snir, and Cynthia A. Patterson. *Getting Up to Speed: The Future of Supercomputing*. Washington, DC: National Academies Press, 2005. Accessed August 8, 2012. research.microsoft.com/en-us/um/people/blampson/72-cstb -supercomputing/72-cstb-supercomputing.pdf.

IBM. "Roadrunner." June 2008. Accessed August 8, 2012. www-03.ibm .com/press/us/en/pressrelease/24405.wss.

Jana, Reena. "Green IT: Corporate Strategies." *Bloomberg Businessweek*, February 11, 2008. Accessed August 8, 2012. www.businessweek.com /stories/2008-02-11/green-it-corporate-strategiesbusinessweek -business-news-stock-market-and-financial-advice.

McMillan, Robert. "Intel Sees Exabucks in Supercomputing's Future." *Wired*, January 24, 2012. Accessed August 8, 2012. www.wired .com/wiredenterprise/2012/01/supercomputings-future.

Museum of American Heritage. "The Technology of Storage." May 6, 2010. Accessed August 8, 2012. www.moah.org/brains/ technology.html.

Peckham, Matt. "The Collapse of Moore's Law: Physicist Says It's Already Happening." *Time*, May 1, 2012. Accessed August 8, 2012. techland.time.com/2012/05/01/the-collapse-of-moores-law -physicist-says-its-already-happening.

Perry, Douglas. "You'd Need a 1 Petaflop to Score #20 Rank in Top 500 List." Tom's Hardware. July 1, 2012. Accessed August 8, 2012. www.tomshardware.com/news/supercomputer-top500 -petaflop-servers-DOE,16042.html.

Yonck, Richard. "The Supercomputer Race, Revisited." World Future Society, June 20, 2011. Accessed August 8, 2012. www.wfs .org/content/supercomputer-race-revisited.

| | | | |

||||| WILL TOUCH SCREENS EVER BE FINGERPRINT-PROOF?

Berman, Art. "Researchers Developing a Smudge Proof Coating." *Display Daily*, December 9, 2011. Accessed September 7, 2012. displaydaily.com/2011/12/09/researchers-developing-a-smudge-proof-coating.

Bonnington, Christina. "Patent App Shows How Apple Makes Touch Displays Fingerprint-Proof." *Wired*, August 12, 2011. Accessed September 7, 2012. www.wired.com/gadgetlab/2011/08/apple-oleophobic-patent.

Humphries, Matthew. "Toray's Touchscreen Film Promises to Cut Down on Fingerprints." Geek.com. February 24, 2012. Accessed September 7, 2012. www.geek.com/articles/mobile/torays-touchscreen-film-promises-to-cut-down-on-fingerprints-20120224.

Kee, Edwin. "Self-Cleaning Paint Might Deliver Smudge-Resistant Touchscreen Displays." Ubergizmo. July 20, 2012. Accessed September 7, 2012, www.ubergizmo.com/2012/07/self-cleaning-paint-might-deliver-smudge-resistant-touchscreen-displays.

Murph, Darren. "Targus Dishes 'Fingerprint-Resistant' Screen Protectors, Diminutive Chargers." Engadget. January 7, 2010. Accessed September 7, 2012. www.engadget.com/2010/01/07/targus-dishes-fingerprint-resistant-screen-protectors-diminut.

Spector, Lincoln. "How to Clean Your Smartphone." *PC World*, July 18, 2011. Accessed September 7, 2012. www.pcworld.com/article/226625/how_to_clean_your_smartphone.html.

| | | | |

||||| HOW WILL COMPUTERS EVOLVE OVER THE NEXT ONE HUNDRED YEARS?

Bone, Simone and Matias Castro. "A Brief History of Quantum Computing."

Imperial College (London), Department of Computing. 1997. www.doc.ic.ac.uk/~nd/surprise_97/journal/vol4/spb3/.

Dubash, Michael. "Moore's Law Is Dead, Says Gordon Moore." *Techworld*, April 13, 2005. Accessed March 17, 2010. news .techworld.com/operating-systems/3477/moores-law-is-dead -says-gordon-moore.

Jet Propulsion Laboratory. "Using 'Nature's Toolbox,' a DNA Computer Solves a Complex Problem." California Institute of Technology. March 14, 2002. Accessed March 18, 2010. www.jpl.nasa.gov /releases/2002/release_2002_63.html.

Massachusetts Institute of Technology. "Center for Extreme Quantum Information Theory, MIT." *TechNews*, March 2007. Accessed March 18, 2010. web.mit.edu/newsoffice/2007/quantum.html.

ScienceDaily. "Optical Computer Closer: Optical Transistor Made from Single Molecule." ETH Zurich. July 3, 2009. Accessed March 17, 2010. www.sciencedaily.com/releases /2009/07/090702080119.htm.

ScienceDaily. "Super-Fast Computers of the Future." Queen's University, Belfast. September 3, 2009. Accessed March 18, 2010. www .sciencedaily.com/releases/2009/09/090901082855.htm.

Venere, Emil. "Purdue Researchers Stretch DNA on Chip, Lay Track for Future Computers." Purdue News Service. October 2003. Accessed March 18, 2010. news.uns.purdue.edu /html4ever/031007.Ivanisevic.DNA.html.

| | | | |

IIIIᵢI HEADS IN THE CLOUDS: HOW CLOUD COMPUTING WORKS

Bogatin, Donna. "Google CEO's New Paradigm: 'Cloud Computing and Advertising Go Hand-in-Hand.'" ZDNet. August 23,

2006. Accessed March 11, 2008. blogs.zdnet.com/micro-markets/?p=369.

Brodkin, Jon. "IBM Unveils 'Cloud Computing.'" *Network World*, November 19, 2007.

Carnegie Mellon Software Engineering Institute. "Middleware." Accessed March 12, 2008. resources.sei.cmu.edu/library.

Carr, Nicholas. "'World Wide Computer' Is on Horizon." *USA Today*, February 25, 2008.

Hickins, Michael. "Cloud Computing Gets Down to Earth." *eWEEK*, January 21, 2008.

IBM. "IBM Introduces Ready-to-Use Cloud Computing," November 15, 2007. Accessed March 12, 2008. www-03.ibm.com/press/us/en/pressrelease/22613.wss.

IT Week. "Report Sees Big Shift in IT Delivery." November 5, 2007.

Lohr, Steve. "Cloud Computing and EMC Deal." *New York Times*, February 25, 2008.

Lohr, Steve. "Google and IBM Join in 'Cloud Computing' Research." *New York Times*, October 8, 2007.

Lohr, Steve. "IBM to Push 'Cloud Computing,' Using Data from Afar." *New York Times*, November 15, 2007.

Markoff, John. "An Internet Critic Who Is Not Shy about Ruffling the Big Names in High Technology." *New York Times*, April 9, 2001.

Markoff, John. "Software via the Internet: Microsoft in 'Cloud' Computing." *New York Times*, September 3, 2007.

McAllister, Neil. "Server Virtualization under the Hood." *InfoWorld*, February 12, 2007. Accessed March 11, 2008. www.infoworld.com/article/07/02/12/07FEvirtualserv_1.html.

Naone, Erica. "Computer in the Cloud." *MIT Technology Review*, September 18, 2007. Accessed March 12, 2008. www.technologyreview.com/Infotech/19397/?a=f.

Network World. "The Future of IT? It's Not All Bad News, Nick Carr Says." January 14, 2008.

Swanson, Bret and George Gilder. "Unleashing the 'Exaflood.'" *Wall Street Journal*, February 22, 2008.

¦ ¦ ¦ ¦

||||¦ PART II: POWER PLAYS: COOL WAYS TO ENERGIZE OUR FUTURE

Bellows, Barbara. "Solar Greenhouse." National Sustainable Agriculture Information Service. 2008. Accessed June 24, 2009. attra.ncat .org/attra-pub/solar-gh.html.

Biello, David. "Sunny Outlook: Can Sunshine Provide All U.S. Electricity?" *Scientific American*, 2007. Accessed June 24, 2009. www.scientificamerican.com/article.cfm?id=sunny-outlook -sunshine-provide-electricity.

Brakmann, Georg, et al. "Concentrated Solar Thermal Power—Now!" European Solar Thermal Power Industry Association, IEA SolarPaces, and Greenpeace International. 2005. Accessed June 24, 2009. www.estelasolar.eu/fileadmin/ESTELAdocs /documents/OPUS_-_Concentrated-Solar-Thermal-Power -Plants-2005.pdf.

GreatHomeImprovements.com. "Thermal Chimneys for Home Cooling." Accessed June 24, 2009. www.greathomeimprovements.com /Nov06theme/housecooling/thermal_chimneys_for_home _cooling.php.

Hobby-Greenhouse.com. "Free Solar Lean-To Greenhouse Plans." Accessed June 24, 2009. www.hobby-greenhouse.com /FreeSolar.html.

Hutchinson, Alex. "Solar Thermal Power May Make Sun-Powered Grid a Reality." *Popular Mechanics*, November 1, 2008. Accessed

June 24, 2009. www.popularmechanics.com/science/research
/4288743.html.

Kanellos, Michael. "Solar Thermal: Which Technology Is Best?"
Greentech Media. April 28, 2009. Accessed June 24, 2009.
www.greentechmedia.com/articles/solar-thermal-which
-technology-is-best-6091.html.

Knier, Gil. "How Do Photovoltaics Work?" NASA. Accessed June 24,
2009. science.nasa.gov/headlines/y2002/solarcells.htm.

LaMonica, Martin. "Solar Thermal Plants Go Back to the Future."
CNET News. September 9, 2007. Accessed June 24, 2009.
news.cnet.com/Solar-thermal-plants-go-back-to-the-future
/2100-11392_3-6206822.html?tag=mncol.

Meisen, Peter and Oliver Pochert. "A Study of Very Large Solar
Desert Systems with the Requirements and Benefits to Those
Nations Having High Solar Irradiation Potential." Global
Energy Network Institute (GENI). July 2006. Accessed June
24, 2009. www.geni.org/globalenergy/library/energytrends
/currentusage/renewable/solar/solar-systems-in-the-desert
/Solar-Systems-in-the-Desert.pdf.

Southface. "How Solar Thermal and Photovoltaics Work." Accessed June
24, 2009. www.southface.org/solar/solar-roadmap/solar_how
-to/solar-how_solar_works.htm.

U.S. Department of Energy. "Solar Energy—Energy from the Sun."
Energy Kid's Page, Energy Information Administration.
Accessed June 24, 2009. www.eia.doe.gov/kids/energyfacts
/sources/renewable/solar.html.

U.S. Department of Energy. "Solar Energy Technologies Program: Dish/
Energy Systems." Energy Efficiency and Renewable Energy.
Accessed June 24, 2009. www1.eere.energy.gov/solar/dish
_engines.html.

U.S. Department of Energy. "Solar Energy Technologies Program: Linear

Concentrator Systems." Energy Efficiency and Renewable
Energy. Accessed June 24, 2009. www1.eere.energy.gov/solar
/linear_concentrators.html.

U.S. Department of Energy. "Solar Energy Technologies Program:
Power Tower Systems." Energy Efficiency and Renewable
Energy. Accessed June 24, 2009. www1.eere.energy.gov/solar
/power_towers.html.

U.S. Department of Energy. "Sunshot Initiative." July 31, 2013. Accessed
March 7, 2014. www1.eere.energy.gov/solar/sunshot/csp
_storage_awards.html.

Williams College. "Sustainability." Accessed June 24, 2009. www
.williams.edu/resources/sustainability/green_buildings/passive
_solar.php?topic=cooling.

World of Solar Thermal. "Low Temperature Collectors." Accessed June
24, 2009. www.worldofsolarthermal.com/vbnews.php?do=vie
warticle&artid=8&title=low-temperature-collector.

| | | | |

CHASING THE SUN: HOW SOLAR THERMAL POWER WORKS

Northwest Power and Conversation Council. "Megawatt." Accessed
March 22, 2013. www.nwcouncil.org/history/megawatt.

| | | | |

HOW SPRAY-ON SOLAR PANELS WORK

Alternative Energy. "Spray-On Solar Panels." February 12, 2009.
Accessed July 14, 2009. www.alternative-energy-news.info
/spray-on-solar-panels.

Biello, David. "Solar Power Lightens Up with Thin-Film Technology."
Scientific American, October 20, 2008. Accessed July 14, 2009.

www.scientificamerican.com/article.cfm?id=solar-power
-lightens-up-with-thin-film-cells.

Douglas, George. "Roadmap to Guide U.S. Photovoltaics Industry in
21st Century." National Renewable Energy Laboratory.
January 20, 2000. Accessed July 14, 2009. www.nrel.gov
/news/press/2000/00200roadmap.html.

Eberspacher, Chris, Karen Pauls, and Jack Serra. "Non-Vacuum
Processing of CIGS Solar Cells." Photovoltaic Specialists
Conference, IEEE. 2002. Accessed July 14, 2009. ieeexplore
.ieee.org/Xplore/login.jsp?url=http%3A%2F%2Fieeexplore
.ieee.org%2Fiel5%2F8468%2F26685%2F01190657
.pdf%3Farnumber%3D1190657&authDecision=-203.

Entrepreneur. "Spray-On Solar Panels for Cost Efficiency." 2009.

Evans, Paul. "Scientists Developing Spray-On Solar Panels." Gizmag.
February 6, 2009. Accessed July 14, 2009. www.gizmag.com
/spray-on-solar-panels/10916.

Goho, Alexandra. "Infrared Vision: New Material May Enhance Plastic
Solar Cells." *Science News*, January 19, 2005.

Gordon, Jacob. "Thin-Film Solar Technology Could Be Seriously
Clobbering Fossil Fuels in Ten Years." TreeHugger. February
20, 2007. Accessed July 14, 2009, www.treehugger.com
/files/2007/02/thinfilm-solar-clobbering-oil.php.

Hall, Kenji. "Japan Pushes toward Solar Energy Future." *Bloomberg
Businessweek*, February 25, 2009. Accessed July 14, 2009.
www.businessweek.com/globalbiz/blog/eyeonasia/archives
/2009/02/japan_pushes_to.html.

Lovgren, Stefan. "Spray-On Solar-Power Cells Are True Breakthrough."
National Geographic, January 14, 2005. Accessed July 14, 2009.
news.nationalgeographic.com/news/2005/01/0114_050114
_solarplastic.html.

Malsch, Ineke. "Thin Films Seek a Solar Future." *The Industrial Physicist*.

Accessed July 14, 2009. www.aip.org/tip/INPHFA/vol-9/iss
-2/p16.html.

Stohr, Stephanie. "Spray-On Solar Panels Developed." *Cosmos Magazine*,
February 3, 2009. Accessed July 14, 2009. www.cosmosmagazine
.com/news/2511/spray-solar-panels-developed.

Weber, Klaus. "Photovoltaic Processes." Australian National University.
sun.anu.edu.au/cells.

Weber, Klaus. "Spray-On Material to Lead to Cheaper Solar Panels."
Australian National University. January 21, 2009. Accessed July
14, 2009. news.anu.edu.au/?p=923.

⊦ ⊦ ⊦ ⊦

⊪ COOL OFF IN THE SUN: HOW SOLAR AIR CONDITIONERS WORK

American Council for an Energy-Efficient Economy. "Air Conditioning."
Accessed July 23, 2009. www.aceee.org/consumer/cooling.

Bhattacharya, Shaoni. "European Heat Wave Caused 35,000 Deaths."
NewScientist, October 10, 2003. Accessed July 23, 2009. www
.newscientist.com/article/dn4259-european-heatwave-caused
-35000-deaths.html.

Bluejay, Michael. "32 Super Tips for Saving Money on Cooling and Air
Conditioning Costs." Accessed July 23, 2009. michaelbluejay
.com/electricity/cooling.html.

Center for Climate and Energy Solutions. "Global Surface Temperature
Trends." Accessed July 23, 2009. www.c2es.org/facts-figures
/trends/surface-temp.

Environmental News Network. "Solar Powered Air Conditioner
Released." August 11, 2008. Accessed July 23, 2009. www
.enn.com/sci-tech/article/37889.

LaMonica, Martin. "Using Solar Energy to Keep Homes Cool." CNET

News. March 20, 2007. Accessed July 23, 2009. news.cnet
.com/2100-11392_3-6168616.html.

Market Wire. "GreenCore Solar Air Conditioners to Be Installed in
McDonald's Restaurant." August 6, 2008. Accessed July 23,
2009. www.reuters.com/article/pressRelease/idUS128609+06
-Aug-2008+MW20080806.

Military & Aerospace Electronics. "U.S. Navy Adopts GreenCore Solar
Air Conditioners." July 30, 2008. Accessed July 23, 2009.
mae.pennnet.com/display_article/335686/32/NEWS/none
/none/1/US-Navy-adopts-GreenCore-solar-air-conditioners.

National Climatic Data Center. "Global Warming Frequently Asked
Questions." Accessed July 23, 2009. www.ncdc.noaa.gov
/cmb-faq/globalwarming.html.

Sierra Solar Systems. "SolCool Hybrid AC/DC Air Conditioning
System." Accessed July 23, 2009. www.sierrasolar.com
/manufacturers.php?manufacturer_id=179.

Solar Tech Times. "2010 Toyota Prius Has New Solar-Power Ventilation
System." 2009.

Toyota. "Toyota Prius." Accessed July 23, 2009. www.toyota.com
/prius/#!/features.

U.S. Department of Energy. "Energy Saver: Heating and Cooling."
Accessed July 23, 2009. www.energysavers.gov/your_home
/space_heating_cooling/index.cfm/mytopic=12360.

U.S. Department of Energy. "Greenhouse Gases, Climate Change, and
Energy." Energy Information Administration. Accessed July 23,
2009. www.eia.gov/oiaf/1605/ggccebro/chapter1.html.

U.S. Environmental Protection Agency. "Future Temperature Changes."
Accessed July 23, 2009. www.epa.gov/climatechange/science
/futuretc.html.

IIIII LITERARY POWER: COULD STEAMPUNK INSPIRE THE FUTURE OF ENERGY?

Atterbury, Paul. "Victorian Technology." BBC. February 17, 2011. Accessed December 2, 2011. www.bbc.co.uk/history/british /victorians/victorian_technology_01.shtml.

Geneva Historical Society. "Clothing in 1840." Accessed December 2, 2011. genevahistoricalsociety.com/wp-content/uploads/2014/03 /clothing.pdf.

Lira, Carl. "Brief History of the Steam Engine." Michigan State University. February 4, 2006. Accessed December 2, 2011. www.egr.msu .edu/~lira/supp/steam.

McFedries, Paul. "Steampunk." Word Spy. Accessed December 2, 2011. www.wordspy.com/words/steampunk.asp.

Mork, Rachel. "How Does Steam Power Work?" Life123. Accessed December 2, 2011. www.life123.com/parenting/education /steam/how-does-steam-power-work.shtml.

National Geographic. "8 Jules Verne Inventions That Came True." February 8, 2011. Accessed December 2, 2011. news.nationalgeographic .com/news/2011/02/pictures/110208-jules-verne-google -doodle-183rd-birthday-anniversary/#/jules-verne-inventions -splash-landing-splashdown-spacecraft_32038_600x450.jpg.

Page, Lewis. "Back to Gaslight, Coal or Steam Power. It's the Future." The Register, June 21, 2011. Accessed December 2, 2011. www .theregister.co.uk/2011/06/21/gaslight_steampunk_fuel_cells.

Rutgers University. "Steam Turbines." Center for Advanced Energy Systems. Accessed December 2, 2011. caes.rutgers.edu/research.

Spirax Sarco. "Steam, the Energy Fluid." Accessed December 2, 2011. www.spiraxsarco.com/resources/steam-engineering-tutorials /introduction/steam-the-energy-fluid.asp.

Sullivan, Chris. "Could Steam Be the New Green Energy?" My Northwest,

August 3, 2011. Accessed December 2, 2011. mynorthwest.
        com/?nid=11&sid=524627.

U.S. Department of Energy. "Steam." Accessed December 2, 2011. www1
        .eere.energy.gov/manufacturing/tech_deployment/steam.html.

U.S. Energy Information Administration. "Electricity in the U.S."
        Accessed December 2, 2011. www.eia.gov/kids/energy
        .cfm?page=electricity_in_the_united_states-basics.

| | | | |

IIIII WHAT ARE WE GASSING ABOUT? HOW GASIFICATION WORKS

Captain, Sean. "Turning Black Coal Green." *Popular Science*, February
        1, 2007. Accessed May 15, 2009. www.popsci.com/scitech
        /article/2007-02/turning-black-coal-green.

Folger, Tim. "Can Coal Come Clean?" *Discover Magazine*, December
        18, 2006. Accessed May 15, 2009. discovermagazine
        .com/2006/dec/clean-coal-technology#.UT9nv6X3Awc.

Gasification Technologies Council. "Gasification: Redefining Clean
        Energy." 2008. Accessed May 15, 2009. ccdtstorage.blob.core
        .windows.net/assets/resources/Final_whitepaper.pdf.

Haq, Zia. "Biomass for Electricity Generation." U.S. Energy Information
        Administration, May 13, 2002. Accessed May 15, 2009. www
        .eia.doe.gov/oiaf/analysispaper/biomass.

Jenkins, Steve. "Gasification 101." Gasification Technologies Workshop.
        May 13, 2008. Accessed May 15, 2009. www.precaution.org
        /lib/igcc101.pdf.

LaFontaine, H. and F.P. Zimmerman. "Construction of a Simplified
        Wood Gas Generator for Fueling Internal Combustion Engines
        in a Petroleum Emergency." Federal Emergency Management

Agency. March 1989. Accessed May 15, 2009. www.woodgas
.net/files/FEMA_emergency_gassifer.pdf.

Oak Ridge National Laboratory. "Biofuels from Switchgrass: Greener
Energy Pastures." Accessed May 15, 2009. bioenergy.ornl.gov
/papers/misc/switgrs.html.

Rajvanshi, Anil K. "Biomass Gasification." Accessed May 15, 2009.
www.nariphaltan.org/nari/pdf_files/gasbook.pdf.

| | | | |

IIIₗI HOLY HYDROGEN: COULD HYDROGEN BE THE FUEL OF THE FUTURE?

U.S. Census Bureau. "U.S. & World Population Clocks." Accessed
March 22, 2013. www.census.gov/main/www/popclock.html.

| | | | |

IIIₗI FARTING MICROBES AND MORE: FIVE EMERGING TECHNOLOGIES IN THE ENERGY INDUSTRY

Allight. "New Mining Technology Promises Clean Energy for 200
Years." Greener Ideal. April 2, 2012. Accessed August 8,
2012.   www.greenerideal.com/alternative-energy/0402-new
-mining-technology-promises-clean-energy-for-200-years.

Biello, David. "New Energy-Dense Battery Could Enable Long-
Distance Electric Cars." Scientific American, February 27,
2012. Accessed August 8, 2012. www.scientificamerican.com
/article.cfm?id=new-energy-dense-battery-could-enable-long
-distance-electric-cars.

Bullis, Kevin. "Microgrid Keeps the Power Local, Cheap, and Reliable."
MIT Technology Review, July 23, 2012. Accessed August 8,

2012. www.technologyreview.com/news/428533/microgrid
-keeps-the-power-local-cheap-and.

Bullis, Kevin. "Using Ozone to Clean Up Fracking." *MIT Technology Review*, August 1, 2012. Accessed August 8, 2012. www
.technologyreview.com/news/428591/using-ozone-to-clean
-up-fracking.

Espinas, Jerico. "New Bladeless Turbine Revolutionizes Wind Industry." Greener Ideal. August 7, 2012. Accessed August 8, 2012. www
.greenerideal.com/alternative-energy/0807-new-bladeless
-turbine-revolutionizes-wind-industry.

FutureGen Alliance. "FutureGen 2.0 Project." Accessed August 8, 2012. www.futuregenalliance.org/futuregen-2-0-project.

LaMonica, Martin. "Tidal Turbine to Take the Plunge in the Bay of Fundy." *MIT Technology Review*, July 25, 2012. Accessed August 8, 2012. www.technologyreview.com/view/428620
/tidal-turbine-to-take-the-plunge-in-the-bay-of.

Saphon Energy. Accessed August 8, 2012. www.saphonenergy.com
/index.php.

ScienceDaily. "Breaking the Barriers for Low-Cost Energy Storage: Battery Could Help Transition to Renewable Energy Sources." University of Southern California. August 1, 2012. Accessed August 8, 2012. www.sciencedaily.com
/releases/2012/08/120801154847.htm.

ScienceDaily. "Scientists Make Microbes to Make 'Clean' Methane." Stanford University. July 27, 2012. Accessed August 8, 2012. www.sciencedaily.com/releases/2012/07/120727144534.htm.

Trabish, Herman K. "CSP and PV Solar Make Each Other More Valuable." GreenTech Solar. April 23, 2012. Accessed August 8, 2012. www.greentechmedia.com/articles/read/CSP-and
-PV-Solar-Make-Each-Other-More-Valuable.

U.S. Department of Energy. "Concentrating Solar Power." October 2011. Accessed August 8, 2012. www.nrel.gov/docs /fy12osti/52478.pdf.

| | | | |

IIIII PART III. HOME IS WHERE THE ROBOT IS: THE INTELLIGENT FUTURE OF OUR HOUSES, HIGHWAYS, AND HOVERCRAFTS

Alchemy Architects. "2012 weeHouse." Accessed May 3, 2012. weehouse.com/index.html#weeHouse.

Brown, Steve. "Builders Expect Home Sizes to Keep Shrinking." RIS Media. January 18, 2011. Accessed May 3, 2012. rismedia.com /2011-01-17/builders-expect-home-sizes-to-keep-shrinking.

Frangos, Alex. "The Green House of the Future." *Wall Street Journal*, April 27, 2009. Accessed May 4, 2012. online.wsj.com/article/ SB124050414436548553.html.

Indiviglio, Daniel. "The Future of Home: Urban and Smaller, but Still Owned." *The Atlantic*, September 20, 2011. Accessed May 3, 2012. www.theatlantic.com/business/archive/2011/09/the-future -of-home-urban-and-smaller-but-still-owned/245394.

Johnston, Ian. "World's Cities to Expand to More Than Twice the Size of Texas by 2030." MSNBC. March 27, 2012. Accessed May 4, 2012. worldnews.msnbc.msn.com /_news/2012/03/27/10887250-worlds-cities-to-expand-by -more-than-twice-the-size-of-texas-by-2030.

Llanos, Miguel. "Could This $30 Million Green Tower Be the Future of World Cities?" MSNBC. March 20, 2012. Accessed May 2, 2012. usnews.msnbc.msn.com/_news/2012/03/20/10226909-could -this-30-million-green-tower-be-the-future-of-world-cities.

Marshall, Jonathan. "Solar Decathlon Shows Homes of the Future Smaller,

Greener." *PGE Currents*, September 28, 2011. Accessed May 3, 2012. www.pgecurrents.com/2011/09/28/solar-decathlon-shows-homes-of-the-future-are-smaller-greener.

Ray, Leah. "Can Super Tall Be Super Green?" GenslerOnCities. November 16, 2010. Accessed May 3, 2012. www.gensleron.com/cities/2010/11/16/can-super-tall-be-super-green.html.

Sellar Group. "Welcome to the Shard." Accessed May 4, 2012. the-shard.com/overview.

University of Maryland. "WaterShed." 2011. Accessed May 3, 2012. solarteam.org/design.

U.S. Department of Energy Solar Decathlon. Accessed May 5, 2012. www.solardecathlon.gov.

| | | | |

IIII REAL-ESTATE REALITY CHECK: WHAT WILL HOMES LOOK LIKE IN FIFTY TO ONE HUNDRED YEARS?

BUILDER. "Concept Home 2010: A Home for the New Economy." Accessed May 23, 2012. www.builderconcepthome2010.com/concept-home.php.

Cofer, Brittany. "Builder Uses Durability, Efficiency to Go Green." *Chattanooga Times Free Press*, November 13, 2010. Accessed May 23, 2012. www.timesfreepress.com/news/2010/nov/13/builder-uses-durability-efficiency-to-go-green.

Fillingham, Nic. "Microsoft Campus Tours: The Microsoft Home." Channel 9. May 2, 2011. Accessed May 23, 2012. channel9.msdn.com/Series/CampusTours/Microsoft-Campus-Tours-The-Microsoft-Home.

Gumbel, Peter. "Building Materials: Cementing the Future." *Time*, December 4, 2008. Accessed May 23, 2012. www.time.com/time/magazine/article/0,9171,1864315,00.html.

Joachim, Mitchell, et al. "Fab Tree Hab." Terreform ONE. Accessed May 23, 2012. www.terreform.org/projects_habitat_fab.html.

Joachim, Mitchell, et al. "In Vitro Meat Habitat." Terreform ONE. Accessed May 23, 2012. www.terreform.org/projects_habitat_meat.html.

Joachim, Mitchell, et al. "Mycoform." Terreform ONE. Accessed May 23, 2012. www.terreform.org/projects_habitat_mycoform.html.

Johnson, Amy. "High-Performance Concrete." *Concrete Decor*, Spring 2008. Accessed May 3, 2012. www.concretedecor.net/decorative concretearticles/counter-culture/spring-2008/high-performance -concrete.

Landscape Online. "Home Sizes Shrink." Accessed May 3, 2012. www .landscapeonline.com/research/article/14264.

Ms. Smith. "Microsoft's Automated Future Home, What Can Go Wrong?" Network World. June 8, 2011. Accessed May 3, 2012. www .networkworld.com/community/blog/microsoft%E2%80%99s -automated-future-home-what-can-go.

Pearson, Ian. "What Do Solar Panels on Your Roof Say about You?" *Futurizon* (blog). April 28, 2012. Accessed May 3, 2012. timeguide.wordpress.com/2012/04/28/what-do-solar-panels -on-your-roof-say-about-you.

Sullivan, Jenny. "Builder's Online Concept Becomes Reality." *BUILDER*, June 17, 2010. Accessed May 3, 2012. www .builderonline.com/design/builders-online-concept-home -becomes-reality.aspx.

U.S. Department of Energy. "Solar Decathlon." Accessed May 3, 2012. www.solardecathlon.gov.

| | | | |

IT KNOWS WHERE YOU LIVE: HOW THE AWARE HOME WORKS

AARP. "What Is Universal Design." Accessed July 30, 2008. www
.aarp.org/families/home_design/universaldesign/a2004-03
-23-whatis_univdesign.html.

Alzheimer's Society. "Safety in the Home." Accessed July 29, 2008.
www.alzheimers.org.uk/factsheet/503.

AutismSpeaks.org. "What Is Autism?" Accessed July 31, 2008. www
.autismspeaks.org/whatisit/index.php.

Aware Home Research Initiative, Georgia Institute of Technology.
Accessed July 28, 2008. awarehome.imtc.gatech.edu.

Baldauf, Sarah. "Taking Care of Your Parents: Preparing the Home."
U.S.News & World Report, November 2, 2007. Accessed July
31, 2008. health.usnews.com/articles/health/2007/11/02/taking
-care-of-your-parents-preparing-the-home.html.

Canada, Carol. "Adapting the Home for Alzheimer's and Dementia
Sufferers." CareGuide@Home. Accessed July 28, 2008.

Centers for Disease Control and Prevention. Chronic Disease Notes &
Reports, June 2, 2007. Accessed July 30, 2008. www.cdc.gov
/aging/pdf/cdnr.june.2007.pdf.

Centers for Disease Control and Prevention. "Chronic Disease Overview."
Accessed July 30, 2008. www.cdc.gov/chronicdisease/overview
/index.htm.

Edmonds, Molly. "How Is an Aging Baby Boomer Generation Changing
the Design of Homes." HowStuffWorks.com. Accessed July 29,
2008. home.howstuffworks.com/baby-boomer-design.htm.

Edmonds, Molly. "How Smart Homes Work." HowStuffWorks.com.
Accessed July 29, 2008. home.howstuffworks.com/smart
-home.htm.

Georgia Institute of Technology. "Overview of the Human Factors &
Aging Laboratory." Accessed July 28, 2008. hfaging.gatech.edu.

Jones, Brian, et al. "Aware Home Research Initiative at the Georgia Institute of Technology." Georgia Institute of Technology. Accessed July 30, 2008. awarehome.imtc.gatech.edu.

Kidd, Cory, et al. "The Aware Home: A Living Laboratory for Ubiquitous Computing Research." Georgia Institute of Technology. Accessed August 1, 2008. www.cc.gatech.edu/fce/ahri /publications/cobuild99_final.PDF.

MedicineNet.com. "Definition of Chronic Disease." Accessed July 30, 2008. www.medterms.com/script/main/art.asp ?articlekey=33490.

National Association of Home Builders. "Aging in Place Remodeling Checklists." Accessed July 28, 2008. www.nahb.org/generic .aspx?sectionID=717&genericContentID=89801.

Partnership for Solutions. "Chronic Conditions: Making the Case for Ongoing Care." Johns Hopkins University and the Robert Wood Johnson Foundation. December 2002. Accessed July 30, 2008. www.partnershipforsolutions.org/DMS/files /chronicbook2002.pdf.

Partnership for Solutions. "Survey Reveals Americans' Concerns about Living with Chronic Conditions and Desire for Elected Officials to Take Action to Improve Care." Johns Hopkins University and the Robert Wood Johnson Foundation. Accessed July 30, 2008. www.partnershipforsolutions.org/statistics/perceptions.html.

Price, Ed, (co-director of the Aware Home Research Initiative at the Georgia Institute of Technology), personal correspondence, August 1, 2008.

Riley, Marcia. "Ubiquitous Computing: An Interesting New Paradigm." Georgia Institute of Technology. Accessed August 1, 2008. project.cyberpunk.ru/idb/ubicomp_interesting_new _paradigm.html.

Sanders, Jane. "Sensing the Subtleties of Everyday Life." Research

Horizons. February 10, 2000. Accessed August 1, 2008. gtresearchnews.gatech.edu/reshor/rh-win00/main.html.

Trevey, John. "Design Techniques for the Homes of Alzheimers Patients." *Disabled World*, August 23, 2007. Accessed July 29, 2008. www .disabled-world.com/artman/publish/article_1516.shtml.

| | | | |

IIIIII NOT JUST HOUSING HYPE: FIVE AWESOME FUTURE HOME TECHNOLOGIES YOU'LL LOVE

Bonnington, Christina. "Apple Patent Describes a More Secure Face Recognition System." *Wired*, May 10, 2012. Accessed September 9, 2012. www.wired.com/gadgetlab/2012/05 /apple-3d-facial-recognition.

Future Technologies. "Lighting Control Benefits & Possibilities." Accessed September 4, 2012. futuretechnologiesinc.com/index/index .php?option=com_content&view=article&id=29&Itemid=26.

Future Technologies. "Track Power." Accessed September 4, 2012. futuretechnologiesinc.com/index/index.php?option=com _content&view=article&id=49&Itemid=38.

Future Technology Portal. "Future Home." Accessed September 4, 2012. www.futuretechnologyportal.com/future-home.htm.

Mapes, Diane. "Are You Ready for the Toilet of the Future?" MSNBC. March 10, 2010. Accessed September 4, 2012. www.msnbc.msn .com/id/35787518/ns/technology_and_science-innovation/t /are-you-ready-toilet-future/#.UEY14NmsRj4.

Ngo, Denise. "Archive Gallery: PopSci Envisions Your Future Home." *Popular Science*, February 25, 2011. Accessed September 4, 2012. www.popsci.com/technology/article/2011-02/archive -gallery-popsci-envisions-your-future-home.

Powerhouse Dynamics. "Residential eMonitor Overview." Accessed

September 9, 2012. www.powerhousedynamics.com/residential
-energy-efficiency.

Schirber, Michael. "How Smart Homes Could Power the Future." Live
Science. July 23, 2008. Accessed September 4, 2012. www
.livescience.com/5019-smart-homes-power-future.html.

ScienceDaily. "New Generation of Home Robots Have Gentle Touch."
Fraunhofer-Gesellschaft. July 14, 2008. Accessed September 4, 2012.
www.sciencedaily.com/releases/2008/07/080710113026.htm.

Seattle City Light. "House of the Near Future Loaded with Energy Saving
Design." *Power Lines*, August 28, 2012. Accessed September
4, 2012. powerlines.seattle.gov/2012/08/28/house-of-the
-near-future-loaded-with-energy-saving-design.

| | | | |

INTELLIGENT TRANSPORTATION SYSTEMS: THE FUTURE OF TRAVEL?

No sources.

| | | | |

FIVE OF THE COOLEST CAR TECHNOLOGIES THAT TRULY HAVE A CHANCE

Bey, Thomas. "Top 10 Future Vehicle Technologies." AskMen. Accessed
December 15, 2011. www.askmen.com/top_10/cars/top-10
-future-vehicle-technology_4.html.

Birch, Stuart. "External Airbag Slows Car in a Crash." *The Telegraph*,
June 11, 2009. Accessed December 15, 2011. www.telegraph
.co.uk/motoring/road-safety/5495705/External-airbag-slows
-car-in-a-crash.html.

BMW. "BMW Augmented Reality." Accessed December 15, 2011.

www.bmw.com/com/en/owners/service/augmented_reality
_introduction_1.html.

Cowen, Tyler. "Can I See Your License, Registration and CPU?" *New York Times*, May 28, 2011. Accessed December 14, 2011. www
.nytimes.com/2011/05/29/business/economy/29view.html.

Green Car Congress. "Ford Showcasing Vehicle-to-Vehicle Communication for Crash Avoidance; Potential for Leveraging WiFi and Smartphones to Extend Quickly the Number of Participating Vehicles." July 19, 2011. Accessed December 14, 2011. www
.greencarcongress.com/2011/07/fordv2v-20110719.html.

Guizzo, Erico. "How Google's Self-Driving Car Works." Discovery News. October 18, 2011. Accessed December 13, 2011. news.discovery.com/autos/how-google-self-driving-car
-works-111018.html.

Kahn, Chris. "Exxon Mobil Predicts Surge in Hybrid Vehicles." *Business Week*, December 8, 2011. Accessed December 16, 2011. www
.businessweek.com/ap/financialnews/D9RGK6C00.htm.

Motoring File. "BMW Group Developing Augmented Reality Window Displays." October 12, 2011. Accessed December 14, 2011. www.motoringfile.com/2011/10/12/bmw-group
-developing-augmented-reality-windshield-displays.

Ramsey, Jonathan. "Mercedes-Benz ESF S400 Concept's 'Braking Bag' in Action." Autoblog. June 14, 2009. Accessed December 15, 2011. www.autoblog.com/2009/06/14/video-mercedes
-benz-esf-s400-concepts-braking-bag-in-action.

Schrank, David, et al. "TTI's 2012 Urban Mobility Report." Texas A&M Transportation Institute, The Texas A&M University System. December 2012. Accessed March 7, 2014. d2dtl5nnlpfr0r.cloud front.net/tti.tamu.edu/documents/mobility-report-2012.pdf.

Thrun, Sebastian. "Leave the Driving to the Car, and Reap Benefits in Safety and Mobility." *New York Times*, December 5, 2011.

Accessed December 12, 2011. www.nytimes.com/2011/12/06
/science/sebastian-thrun-self-driving-cars-can-save-lives-and
-parking-spaces.html.

Truong, Alice. "Do Driverless Cars Offer Safer and More Efficient
Transportation?" Discovery News. October 12, 2011. Accessed
December 13, 2011. dsc.discovery.com/cars-bikes/do-driverless
-cars-offer-safer-and-more-efficient-transportation.html.

Volvo. "Tomorrow's Volvo Car: Body Panels Serve as the Car's
Battery." September 24, 2010. Accessed December 15, 2011.
www.media.volvocars.com/global/enhanced/en-gb/Media
/Preview.aspx?mediaid=35026.

Yvkoff, Liane. "Toyota Demos Augmented-Reality-Enhanced Car
Windows." CNET. July 21, 2011. Accessed December 15,
2011.   reviews.cnet.com/8301-13746_7-20081402-48/toyota
-demos-augmented-reality-enha.

| | | | |

SMART STREETS WITH STREET SMARTS: HOW INTELLIGENT HIGHWAYS WILL WORK

"Cell Internet Use 2012." Pew Internet & American Life Project. Accessed
March 22, 2013. www.pewinternet.org/Reports/2012/Cell
-Internet-Use-2012.aspx.

"Terrafugia." Accessed March 22, 2013. www.terrafugia.com.

| | | | |

CHARIOTS OF HIGHER: HOW SOLAR AIRCRAFT WORK

AFP. "Solar Plane Sets Distance Record on US Tour." Phys.org. May 23,
2013. Accessed March 7, 2014. phys.org/news/2013-05-solar
-plane-distance.html.

Boeing. "Jet Prices." Accessed June 4, 2009. www.boeing.com
/commercial/prices.

Boeing. "Technical Characteristics: Boeing 747-400." Accessed June 4,
2009. www.boeing.com/commercial/747family/pf/pf_400_prod
.html.

Bush, Steven. "Inside QinetiQ's Zephyr Solar Powered Plane." *Electronics
Weekly*, September 28, 2007. Accessed June 4, 2009. www
.electronicsweekly.com/Articles/2007/09/28/42280/inside
-qinetiqs-zephyr-solar-powered-plane.htm.

Del Frate, John, personal interview, June 4, 2009.

Intergovernmental Panel on Climate Change. "What Are the Current
and Future Impacts of Subsonic Aviation on Radiative Forcing
and UV Radiation?" *IPCC Special Report on Aviation and the
Global Atmosphere*, 1999. Accessed June 4, 2009. www.ipcc
.ch/ipccreports/sres/aviation/006.htm#spmfig2a.

Lockheed Martin. "F-22 Raptor." Accessed June 4, 2009. www
.lockheedmartin.com/us/products/f22.html.

National Aeronautics and Space Administration. "Helios Prototype."
Accessed June 4, 2009. www.nasa.gov/centers/dryden/news
/FactSheets/FS-068-DFRC.html.

Noll, Thomas, et al. "Investigation of the Helios Prototype Aircraft
Mishap." January 2004. Accessed June 4, 2009. www.nasa
.gov/pdf/64317main_helios.pdf.

QinetiQ. "QinetiQ's Zephyr UAV Flies for Three and a Half Days to
Set Unofficial World Record for Longest Duration Unmanned
Flight." August 24, 2008. Accessed June 4, 2009. www.qinetiq
.com/home/newsroom/news_releases_homepage/2008/3rd
_quarter/qinetiq_s_zephyr_uav.html.

Solar Impulse. Accessed June 4, 2009. www.solarimpulse.com.

U.S. Department of Defense Advanced Research Projects Agency.
"Vulture Program: Broad Agency Announcement (BAA)

Solicitation 07-51." July 31, 2007. Accessed June 4, 2009. www.findthatpdf.com/search-266810-hPDF/download -documents-20070917-17darpa-vulture.pdf.htm.

U.S. Environmental Protection Agency. "Regulatory Announcement: New Emission Standards for New Commercial Aircraft Engines." Accessed June 4, 2009. nepis.epa.gov/Adobe/PDF /P1001YUC.pdf.

Wachter, Sarah. "Airports Struggle to Reduce Their Pollution." *New York Times*, June 4, 2008. Accessed June 4, 2009. www .nytimes.com/2008/06/04/business/worldbusiness/04iht -rbogair.4.13465825.html?_r=2&scp=9&sq=airport%20 pollution&st=cse.

| | | | | |

FABRIC, FABRIC, EVERYWHERE! HOW FABRIC DISPLAYS WORK

Adwalkers. Accessed November 30, 2007. www.adwalkers.com.

Cambridge Display Technology. Accessed November 30, 2007. www .cdtltd.co.uk/faqs/50.asp.

Collins, Clayton. "'Billboards' That Walk, Talk, And Even Flirt a Little." *The Christian Science Monitor*, July 8, 2004. Accessed November 30, 2007. www.csmonitor.com/2004/0708/p11s01 -wmgn.html.

Elwire.com. "What Is EL Wire?" Accessed November 30, 2007. www .elwires.com/faq.

Guglielmi, Michel and Hanne-Louise Johannesen. "Thermochromic Ink Applied to Wearables." Accessed November 30, 2007. www .diffus.dk/publi-conf/paper_ny.pdf.

Lacasse, K. and W. Baumann. *Textile Chemicals: Environmental Data and Facts*. New York: Springer, 2004.

Philips. "Philips Illuminates IFA 2006 with Production-Ready Lumalive Textile Garments." Press release. August 24, 2006. Accessed November 30, 2007. www.research.philips.com/newcenter /archive/2006/060901-lumalive.html.

T-Shirt TV. Accessed November 30, 2007. www.t-shirttv.com.

U.S. Patent and Trademark Office. "Fabric Display." U.S. Patent Application 20070076407. April 5, 2007. Accessed November 30, 2007. appft1.uspto.gov/netacgi/nph-Parser?Sect1=PTO1& Sect2=HITOFF&d=PG01&p=1&u=%2Fnetahtml%2FPTO% 2Fsrchnum.html&r=1&f=G&l=50&s1=%2220070076407%22 .PGNR.&OS=DN/20070076407&RS=DN/20070076407.

| | | | |

⁞⁞⁞⁞ BACK TO THE FUTURE: FIVE FUTURE TECHNOLOGY MYTHS

Ater, Tal. "NASA Abandons Flying Cars for Greener Flight with a $1.5m Prize for Green Plane Innovation." Green Prophet. August 3, 2009. Accessed August 17, 2009. greenprophet .com/2009/08/03/11134/nasa-abandons-flying-cars-for -green-flight/.

Borenstein, Seth. "Scientists Say Global Warming Inevitable, But Disasters Aren't." Associated Press in the *Seattle Times*, April 3, 2006. Accessed August 17, 2009. seattletimes.nwsource.com/html /nationworld/2002906901_warming03.html.

Brain, Marshall. "Robotic Nation." Accessed August 17, 2009. www .marshallbrain.com/robotic-nation.htm.

Gizmo Watch. "Top Ten Flying Cars." March 22, 2007. Accessed August 17, 2009. www.gizmowatch.com/entry/top-ten-flying-cars.

Hassler, Susan. "Un-Assuming the Singularity." *IEEE Spectrum*, June 2008. Accessed August 17, 2009. www.spectrum.ieee.org /computing/hardware/unassuming-the-singularity.

Hertsgaard, Mark. "It's Much Too Late to Sweat Global Warming." *San Francisco Chronicle*, February 13, 2005. Accessed August 17, 2009, www.sfgate.com/cgi-bin/article.cgi?f=/c/a/2005/02/13/INGP4B7GC91.DTL.

Hickins, Michael. Moore's Law Reaching Statute of Limitations. CBS News. July 22, 2009. Accessed August 17, 2009. www.cbsnews.com/news/moores-law-reaching-statute-of-limitations/.

IEEE Spectrum. "Tech Luminaries Address Singularity." June 2008. Accessed August 17, 2009. www.spectrum.ieee.org/computing/hardware/tech-luminaries-address-singularity.

iSuppli. "Is Moore's Law Becoming Academic?" June 16, 2009. Accessed August 17, 2009. www.isuppli.com/semiconductor-value-chain/news/pages/is-moore-s-law-becoming-academic.aspx.

Joy, Bill. "Why the Future Doesn't Need Us." *Wired*. Accessed August 17, 2009. www.wired.com/wired/archive/8.04/joy_pr.html.

Markoff, John. "Scientists Worry Machine May Outsmart Man." *New York Times*, July 25, 2009. Accessed August 17, 2009. www.nytimes.com/2009/07/26/science/26robot.html.

Markoff, John. "Software That Cares." *New York Times*, July 28, 2009. Accessed August 17, 2009. tierneylab.blogs.nytimes.com/2009/07/28/software-that-cares.

Nuttall, Chris. "Moore's Law Reaches Its Economic Limits." *Financial Times*, July 20, 2009. Accessed August 17, 2009. www.ft.com/intl/cms/s/2/01c8b5aa-754e-11de-9ed5-00144feabdc0.html#axzz2NL0ff358.

O'Keefe, Brian. "The Smartest (or Nuttiest) Futurist on Earth." CNN. May 2, 2007. Accessed August 17, 2009. money.cnn.com/magazines/fortune/fortune_archive/2007/05/14/100008848.

Oxfam. "Suffering the Science: Climate Change, People, and Poverty." July 6, 2009. Accessed August 17, 2009. www.oxfam.org/sites/www.oxfam.org/files/bp130-suffering-the-science.pdf.

Page, Lewis. "DARPA at Work on 'Transformer X,' a Proper Flying Car." *The Register*, May 26, 2009. Accessed August 17, 2009. www.theregister.co.uk/2009/05/26/darpa_flying_car_transformer.

Popplewell, Brett. "Flying Car Not Far Away." *The Chronicle Herald*, July 28, 2009.

Ross, Greg. "An Interview with Douglas R. Hofstadter." *American Scientist*. Accessed August 17, 2009. www.americanscientist.org/bookshelf/pub/douglas-r-hofstadter.

CONTRIBUTORS

Talal Al-Khatib

Kevin Bonsor

Terri Briseno

Jacob Clifton

Laurie L. Dove

Molly Edmonds

Shanna Freeman

Kristen Hall-Geisler

William Harris

Kate Kershner

Patrick J. Kiger

Robert Lamb

Susan L. Nasr

Christopher Neiger

Melanie Radzicki-McManus

Jacob Silverman

Jonathan Strickland

Jessika Toothman

Maria Trimarchi

ABOUT HOWSTUFFWORKS

HowStuffWorks.com is an award-winning digital source of credible, unbiased, and easy-to-understand explanations of how the world actually works. Founded in 1998, the site is now an online resource for millions of people of all ages. From car engines to search engines, from cell phones to stem cells, and thousands of subjects in between, HowStuffWorks.com has it covered. In addition to comprehensive articles, our helpful graphics and informative videos walk you through every topic clearly and objectively. Our premise is simple: demystify the world and do it in a clear-cut way that anyone can understand.

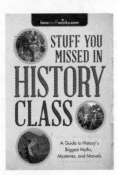

THE SCIENCE OF SUPERHEROES AND SPACE WARRIORS: LIGHTSABERS, BATMOBILES, KRYPTONITE, AND MORE!

DO YOU HAVE WHAT IT TAKES TO BE A SUPERHERO?

You picked out your superpower years ago. You can change into your costume in seconds. You could take out a Sith Lord with your lightning-quick lightsaber moves. Not so fast! Before you can start vanquishing bad guys, it's important to be schooled in the science of saving the world. Come learn the science behind your favorite superheroes and supervillains and their ultracool devices and weapons—from Batmobiles and warp speed to lightsabers, Death Stars, and kryptonite—and explore other cool technologies from the science fiction realm in this dynamic book. Discover:

- How Batman and the Batmobile really work
- 10 *Star Trek* technologies that actually came true
- Whether warp speed and lightsabers are really possible
- If Superman would win against Harry Potter, Sith Lords, and even Chuck Norris!
- How new liquid body armor can make us superhuman
- And more!

Prepare to do battle with the world's most evil masterminds!

THE REAL SCIENCE OF SEX APPEAL: WHY WE LOVE, LUST, AND LONG FOR EACH OTHER

EVER WONDER WHY LOVE MAKES US SO CRAZY? COME DIVE INTO THE *REAL* SCIENCE BEHIND SEX APPEAL AND WHY WE LOVE, LUST, AND LONG FOR EACH OTHER.

Did you know your walk, your scent, and even the food you eat can make you sexier? Or that there are scientifically proven ways to become more successful at dating, especially online? The team at the award-winning website HowStuffWorks reveals the steamy science of love and sex, from flirting to falling in love and everything in between. Discover:

- How aphrodisiacs and sex appeal work (and how to increase yours!)
- Whether love at first sight is scientifically possible
- Why breakup songs hurt so good
- What happens in the brain during an orgasm
- The crazy chemistry behind long-term relationships
- The dope on dating and matchmaking
- And much more!

This dynamic book will show you what to expect—and what to do—the next time someone sets your heart racing.